41st Vietnam National Conference on Theoretical Physics (NCTP-41)

Journal of Physics: Conference Series Volume 865

Nha Trang, Vietnam
1 - 4 August 2016

Editors:

Trinh Xuan Hoang
Hoang Anh Tuan
Vu Ngoc Tuoc

ISBN: 978-1-5108-4296-0
ISSN: 1742-6588

Printed from e-media with permission by:

Curran Associates, Inc.
57 Morehouse Lane
Red Hook, NY 12571

Some format issues inherent in the e-media version may also appear in this print version.

Printed by Curran Associates, Inc. (2017)

For permission requests, please contact the Institute of Physics
at the address below.

Institute of Physics
Dirac House, Temple Back
Bristol BS1 6BE UK

Phone: 44 1 17 929 7481
Fax: 44 1 17 920 0979

techtracking@iop.org

Additional copies of this publication are available from:

Curran Associates, Inc.
57 Morehouse Lane
Red Hook, NY 12571 USA
Phone: 845-758-0400
Fax: 845-758-2633
Email: curran@proceedings.com
Web: www.proceedings.com

Table of contents

Volume 865

41st Vietnam National Conference on Theoretical Physics

1–4 August 2016, Nha Trang, Vietnam

Accepted papers received: 19 June 2017
Published online: 26 June 2017

Preface

Papers

IOP Conf. Series: Journal of Physics: Conf. Series **865** (2017) 011001 doi:10.1088/1742-6596/865/1/011001

Preface

The 41st Vietnam National Conference on Theoretical Physics (NCTP-41) was held during 1-4 August 2016 in Nha Trang, Vietnam.

The NCTP-41 was organized by the Institute of Physics, Vietnam Academy of Science and Technology (IOP-VAST) under the support of the Vietnamese Theoretical Physics Society (VTPS).

This meeting belongs to a series of annual theoretical physics conferences that started in 1976.

The conference has covered a wide range of theoretical physics topics from 4 major fields:

- Particle, nuclear and astro- physics,
- Molecular physics, quantum optics and quantum computation,
- Condensed matter physics,
- Soft matter, biological and interdisciplinary physics.

115 participants have participated in the conference. 2 invited talks, 22 oral and 75 poster contributions were presented.

This volume contains selected papers contributed by the participants.

Editors of the NCTP-41 Proceedings

Trinh Xuan Hoang, Hoang Anh Tuan and Vu Ngoc Tuoc

IOP Conf. Series: Journal of Physics: Conf. Series **865** (2017) 011001 doi:10.1088/1742-6596/865/1/011001

Organizer

- Institute of Physics, Vietnam Academy of Science and Technology (IOP-VAST)

Sponsor

- Vietnam Academy of Science and Technology (VAST)

Honorary Chair

- Nguyen Van Hieu (VAST)

Chair

- Nguyen Ai Viet (IOP-VAST)

Organizing Committee

- Trinh Xuan Hoang (IOP-VAST), Chair
- Bach Thanh Cong (VNU University of Science, Hanoi)
- Hoang Dzung (Vietnam National University, Ho Chi Minh City)
- Le Van Hoang (Ho Chi Minh City Pedagogical University)
- Nguyen Tri Lan (IOP-VAST)

Program Committee

- Hoang Anh Tuan (IOP-VAST), Chair
- Phung Van Dong (IOP-VAST)
- Ho Trung Dung (Ho Chi Minh city Institute of Physics)
- Nguyen The Toan (VNU University of Science, Hanoi)
- Vu Ngoc Tuoc (Hanoi University of Science and Technology)

Proceedings

- Trinh Xuan Hoang (IOP-VAST)
- Hoang Anh Tuan (IOP-VAST)
- Vu Ngoc Tuoc (Hanoi University of Science and Technology)

IOP Conf. Series: Journal of Physics: Conf. Series **865** (2017) 011001 doi:10.1088/1742-6596/865/1/011001

List of Participants

Đào Thu Hằng
Faculty of Physics, Hanoi University of Sciences, Vietnam National University
Email: daohangkhtn@gmail.com

Đặng Trung Sĩ
NCS Viện Vật Lý
Email: dangtrungsi@gmail.com

Bach Thanh Cong
Faculty of Physics, VNU University of science
Email: congbt@vnu.edu.vn

Ca Nguyễn Anh Khoa
Đại học Cần Thơ
Email: khoa.cn1984@gmail.com

Chu Thuy Anh
Institute of Physics
Email: ctanh@iop.vast.vn

Dang Dinh Long
VNU University of Engineering and Technology
Email: ddlong@vnu.edu.vn

Dao Vong Duc
Institute of Physics, VAST
Email: dvduc@iop.vast.vn

Dinh Nguyen Dinh
Institute of Physics, Hanoi
Email: dndinh@iop.vast.ac.vn

Do Son
Ho Chi Minh City University of Technology
Email: dnson@hcmut.edu.vn

Do Chi Nghia
Hanoi Pedagogical University No.2
Email: dochinghia.sp2@moet.edu.vn

Do Q. Tuan
Vietnam National University, Hanoi
Email: tuanqdo@vnu.edu.vn

Do Thi Hong Hai
Department of physics, Hanoi University of Mining and Geology
Email: dohonghaimdc@gmail.com

Do Tuan Long
Faculty of Physics, Hanoi University of Science, Vietnam National University

IOP Conf. Series: Journal of Physics: Conf. Series **865** (2017) 011001 doi:10.1088/1742-6596/865/1/011001

Email: dotuanlong@hus.edu.vn

Doan Quoc Khoa
Quang Tri Teacher Training College
Email: khoa_dqspqt@yahoo.com

Duong Thi Ha
Thai nguyen University of Education
Email: duongha.sp@gmail.com

Duong Thi Man
Institute of Physics, VAST
Email: duongthiman28@gmail.com

Duong Van Loi
Institute of Physics, VAST
Email: loidvtb@gmail.com

Giap Thuy Trang
Thai Nguyen University of Education
Email: giapthuytrang@gmail.com

Ho Khac Hieu
Duy Tan University
Email: hieuhk@gmail.com

Hoang Anh-Tuan
Institute of Physics - VAST
Email: hatuan@iop.vast.ac.vn

Huỳnh Vĩnh Phúc
Trường Đại học Đồng Tháp
Email: hvphuc@dthu.edu.vn

Huong Thi Quynh Le
University of Khanh Hoa
Email: lethiquynhhuong@yahoo.com

Lê Đình
Hue University - College of Education
Email: dinhle52@gmail.com

Lê Đại Nam
Ton Duc Thang University
Email: ldn28593@gmail.com

Lê Đức Thiện
Hoang Van Thu High School for Gitted Students, Thinh Lang Precinct, Hoa Binh City, Hoa Binh Province, Viet Nam
Email: leducthienhvt@gmail.com

Lê Văn Hoàng
Department of Physics, Ho Chi Minh City University of Pedagogy

Email: hoanglv2005@gmail.com

Le Tuan
Hanoi University of Science and Technology
Email: tuan.le@hust.edu.vn

Le Duc Anh
Hanoi National University of Education
Email: anhld@hnue.edu.vn

Le Duy Manh
Institute of Physics, VAST
Email: ldmanh02468@gmail.com

Le Thi Cam Tu
Sai Gon University
Email: letu106@gmail.com

Le Tho Hue
Institute of Physics, VAST
Email: lthue@iop.vast.ac.vn

Le Van Vinh
Department of Computational Physics, Hanoi University of Science and Technology, Vietnam
Email: vinh.levan@hust.edu.vn

Le Van Xuan
Dong Xuan, Soc Son, Ha Noi, Viet Nam
Email: xuanlv.hn@gmail.com

Lương Thị Thêu
ĐHSP Hà Nội 2
Email: luongtheu@gmail.com

Lương Văn Tùng
Đại học Đồng Tháp
Email: lvt1961@yahoo.com.vn

Mai Thi Lan
Hanoi University of Sciences and Technology
Email: lan.maithi@hust.edu.vn

Mẫn Văn Ngữ
Hung Yen Industrial Colleage
Email: nguqlkt@yahoo.com

Ngô Gia Vịnh
Bac Ninh Department of Education and Training
Email: ngogiavinh@bacninh.edu.vn

Nguyễn Đức Hiền
Sở GD & ĐT Gia Lai
Email: n.duchien@gmail.com

IOP Conf. Series: Journal of Physics: Conf. Series **865** (2017) 011001 doi:10.1088/1742-6596/865/1/011001

Nguyễn Huy Thảo
Hanoi Pedagogical University 2
Email: abcthao@gmail.com

Nguyễn Minh Hoa
Hue University of Medicine and Pharmacy
Email: minhhoa2806@gmail.com

Nguyễn Phương Duy Anh
Thủ Dầu Một University
Email: nguyenphuongduyanh@yahoo.com

Nguyễn Từ Niệm
VNU University of Science
Email: nguyentuniem@gmail.com

Nguyễn Thị Bích Hà
Vietnam Academy of Science and Technology
Email: bichha@iop.vast.vn

Nguyễn Thị Hòa
University of Transport and Communications
Email: nthoaly@utc.edu.vn

Nguyễn Văn Hồng
Bộ môn Vật lý Tin học, Viện Vật lý Kỹ Thuật, ĐH Bách Khoa Hà Nội
Email: hong.nguyenvan@hust.edu.vn

Nguyễn Văn Hiệu
Vietnam Academy of Science and Technology
Email: nvhieu@iop.vast.vn

Nguyễn Văn Hoa
Trường Đại học Sư phạm Tp. HCM
Email: nvhoa62@yahoo.com

Nguyen Ai Viet
Institute of Physics
Email: vieta@iop.vast.ac.vn

Nguyen Anh Sang
Hanoi Pedagogical University No.2
Email: anhsangk13@gmail.com

Nguyen Dinh Hien
Center for Theoretical and Computational Physics, Hue University's College of Education
Email: dinhhienphysics@gmail.com

Nguyen Dung Chinh
Theoretical Physics Research Group, Faculty of Applied Sciences, Ton Duc Thang University
Email: ndchinh88@gmail.com

Nguyen Duong Bo
Trung Tâm Vật Lý Lý Thuyết - Viện Vật Lý
Email: nguyenduongbo80@gmail.com

Nguyen Duy Huy
VNU University of Science
Email: huydn@hus.edu.vn

Nguyen Hoang Giang
Comptutational Physics Lab, Institute of Technology, Vietnam National University - Ho Chi Minh City
Email: nhgiang150490@gmail.com

Nguyen Hoang Tung
Department of Nuclear Physics, University of Science, Ho Chi Minh City
Email: nguyen.hoang.tung.13122@gmail.com

Nguyen Hong Son
Trade Union University
Email: atvsld@gmail.com

Nguyen Ngoc Hieu
Duy Tan University
Email: hieunguyenvly@yahoo.com

Nguyen Quang Bau
Faculty of Physics, Hanoi University of Science, Vietnam National University
Email: nguyenquangbau54@gmail.com

Nguyen Thi Dung
Hong Duc University
Email: nguyenthidungvatly@hdu.edu.vn

Nguyen Thi Huong
Thuy Loi University
Email: huonghnue@gmail.com

Nguyen Thi Thao
Faculty of Physics, Hanoi National University of Education
Email: ntthao.hnue@gmail.com

Nguyen Thi Ha Loan
Ha Noi Pedagogical University No. 2
Email: hloansp2@gmail.com

Nguyen Thi Hai Yen
Institute of Physics
Email: nhyen@iop.vast.vn

Nguyen Thi Kim Hue
Can Tho University
Email: nguyenthikimhue0509@gmail.com

IOP Conf. Series: Journal of Physics: Conf. Series **865** (2017) 011001 doi:10.1088/1742-6596/865/1/011001

Nguyen Thi Kim Quyen
Can Tho University
Email: kimquyen929@gmail.com

Nguyen Thi Kim Thanh
Institute of Physics, VAST
Email: nkthanh@iop.vast.vn

Nguyen Thi Phuong Lan
Ha Noi Pedagogical University No. 2
Email: lanntphuong@gmail.com

Nguyen Thi Thanh Ha
Department of Computational Physics, Hanoi University of Science and Technology, Vietnam
Email: ha.nguyenthithanh1@hust.edu.vn

Nguyen Thi Thuy Hang
Ho Chi Minh City University of Technology
Email: hangthuynguyen2001@yahoo.com

Nguyen Thu Nhan
Department of Computational Physics, Hanoi University of Science and Technology, Vietnam
Email: thunhaniep@gmail.com

Nguyen To Nga
PetroVietnam University
Email: jacqueline.nga.nguyen@gmail.com

Nguyen Toan Thang
Institute of Physics, VAST
Email: ntthang@iop.vast.vn

Nguyen Tri Lan
Institute of Physics
Email: ntlan@iop.vast.vn

Nguyen Van Hinh
Ha Noi University of industry
Email: nguyenvanhinh78@gmail.com

Nguyen Van Hop
Department of Physics, Hanoi National University of Education
Email: hopnv@hnue.edu.vn

Nguyen Van Nghia
Thuy Loi University
Email: nghiangv@yahoo.com

Nguyen Van Yen
Hanoi University of Science and Technology
Email: trungyen2512@gmail.com

Nguyen Viet Duc

IOP Conf. Series: Journal of Physics: Conf. Series **865** (2017) 011001 doi:10.1088/1742-6596/865/1/011001

Faculty of Physics, VNU University of science
Email: ducnv84@gmail.com

Nguyen Ba Hung
Vietnam Military Medical University
Email: hungnb80@yahoo.com.vn

Nguyen Ba Vu Chinh
Danang University of Education
Email: vuchinhufo86@gmail.com

O My Na
Ho Chi Minh City University of Technology
Email: myna1707@gmail.com

Pham Khac Hung
Department of Computational Physics, Hanoi University of Science and Technology, Vietnam
Email: hung.phamkhac@hust.edu.vn

Pham Nguyen Thanh Vinh
Ho Chi Minh University of Pedagogy
Email: vinhpham@hcmup.edu.vn

Pham Thanh Thuy
Can Tho University
Email: ptthuy977@gmail.com

Pham Thi Thanh Nga
Water Resources University
Email: nga_ptt@tlu.edu.vn

Pham Tuan Vinh
Dong Thap University
Email: tuanvinhdhdt@gmail.com

Phan Ngọc Hưng
HCMC University of Pedagogy
Email: hungpn@hcmup.edu.vn

Phan Thanh Hùng
Tra Vinh University
Email: pthung.tvu@gmail.com

Phan Thi Ngoc Loan
Ho Chi Minh City University of Pedagogy
Email: loan2vietnam@gmail.com

Phung Duy Khuong
Institute of Physics
Email: pdkhuong@iop.vast.vn

Tran Cong Phong
Vietnam Institute of Educational Sciences

Email: tcphong@moet.edu.vn

Tran Hoai Nam
Duy Tan University
Email: tranhoainamk3@gmail.com

Tran Minh Tien
Thu Dau Mot University
Email: tientm@tdmu.edu.vn

Tran Thi Nhan
Đại học Công Nghiệp Hà Nội
Email: trannhan09@gmail.com

Tran Thi Thu Trang
Halong University
Email: trangtranthithu.c17@moet.edu.vn

Tran Thuy Duong
Hanoi University of Science and Technology
Email: duong.tranthuy@hust.edu.vn

Tran Van Quang
Department of Physics, University of Transport and Communications and Institute of Research and Development, Duy Tan University
Email: tkuangv@gmail.com

Trần Minh Tiến
Viện Vật lý, Viện Hàn lâm Khoa học và Công nghệ Việt Nam
Email: minhtien@iop.vast.vn

Trần Thanh Dũng
Thu Dau Mot University
Email: ttdung24@gmail.com

Trần Đình Thám
Trường ĐH Phạm Văn Đồng
Email: tdtham@pdu.edu.vn

Trần Lan Phương
Ho Chi Minh City University of Pedagogy
Email: tranlanphuongsp@gmail.com

Trần Minh Hiến
Viện Vật lý Tp. HCM
Email: tranminhhien@gmail.com

Trần Thanh Tuyền
Bắc Ninh
Email: tranthanhtuyenqv3@bacninh.edu.vn

Trần Thành Nhật
Institute of Northwest Education and Research

IOP Conf. Series: Journal of Physics: Conf. Series **865** (2017) 011001 doi:10.1088/1742-6596/865/1/011001

Email: tranthnhat@gmail.com

Trần Thị Hải
Faculty of Natural Science, Hong Duc University, Thanh Hoa
Email: tranthihai@hdu.edu.vn

Trịnh Xuân Hoàng
Institute of Physics, VAST
Email: hoang@iop.vast.ac.vn

Truong Dang Hoai Thu
Ho Chi Minh University of Pedagogy
Email: tdhoaithu@gmail.com

Trương Minh Đức
Đại học Sư Phạm Huế
Email: tmduc2009@gmail.com

Trương Trọng Thúc
NCS Viện Vật Lý
Email: truongtrongthuck17@gmail.com

Vinh Dinh Quang
Hanoi National University of Education
Email: vinhhitech@gmail.com

Vo Van Vien
Department of Physics, Tay Nguyen University
Email: wvienk16@gmail.com

Vu Ba Dung
Hanoi University of Mining and Geology
Email: vubazung305@gmail.com

Vu Ngoc Tuoc
Hanoi University of Science and Technology
Email: tuoc.vungoc@hust.edu.vn

Vu Thanh Tra
Can Tho University
Email: vttra@ctu.edu.vn

Peer review statement

All papers published in this volume of *Journal of Physics: Conference Series* have been peer reviewed through processes administered by the proceedings Editors. Reviews were conducted by expert referees to the professional and scientific standards expected of a proceedings journal published by IOP Publishing.

IOP Conf. Series: Journal of Physics: Conf. Series **865** (2017) 012001 doi:10.1088/1742-6596/865/1/012001

On five-dimensional massive (bi-)gravity

Tuan Q Do

Faculty of Physics, VNU University of Science, Vietnam National University, Hanoi 120000, Vietnam

E-mail: tuanqdo@vnu.edu.vn

Abstract. Main results of our recent investigations on five-dimensional scenarios of massive (bi-)gravity will be summarized in this article. In particular, we will show how to construct higher dimensional massive graviton terms from the characteristic equation of square matrix, which is a consequence of the Cayley-Hamilton theorem. Then, we will show whether massive graviton terms of five-dimensional massive (bi-)gravity behave as effective cosmological constants for a number of physical metrics compatible with fiducial ones such as the Friedmann-Lemaitre-Robertson-Walker, Bianchi type I, and Schwarzschild-Tangherlini metrics. Finally, we will show the corresponding cosmological solutions for the five-dimensional massive (bi-)gravity.

1. Introduction

Massive gravity, in which a graviton is assumed to have tiny but non-vanishing mass, has a rich and long story since a seminal paper by Fierz and Pauli (FP) [1]. In particular, it was shown by van Dam and Veltman [2] and Zakharov [3] that the FP theory cannot recover the general relativity by Einstein in the massless limit. However, Vainshtein showed in [4] that non-linear extensions of FP theory might resolve the so-called van Dam-Veltman-Zakharov (vDVZ) discontinuity problem. Unfortunately, Boulware and Deser (BD) claimed, shortly after the paper of Vainshtein, in [5] that a unexpected ghost might arise from the sixth mode of graviton in non-linear FP theory. Therefore, building a non-linear but ghost-free massive gravity has been a great challenge to physicists for quite a long time. Indeed, many attempts to build up a ghost-free non-linear massive gravity have been proposed, e.g., in [6]. However, most of them have not turned out to be successful approaches except a non-linear massive gravity constructed by de Rham, Gabadadze, and Tolley (dRGT) [7] as a generalization of FP massive gravity. As a result, the most important property of the dRGT theory is that it has been proved to be free of the BD ghost by different approaches whatever the form of reference (or fiducial) metric [8].

Consequently, many cosmological and physical aspects of the dRGT theory have been investigated extensively. For example, the dRGT theory has been expected to provide us an alternative solution to the cosmological constant problem associated with the expanding of our universe. Black holes such as the Schwarzschild, Kerr, and charged black holes along with anisotropic metrics such as the Bianchi type I metric have also been investigated in the context of dRGT theory. More interestingly, the massive gravity and condensed matter theories have met each other in the so-called holographic massive gravity. For more detailed information, one can read a recent interesting review paper by de Rham [9].

Along with the above investigations, many extensions of the non-linear massive gravity have also been carried out [9]. One of them is the massive bi-gravity (or bi-metric gravity) proposed

IOP Conf. Series: Journal of Physics: Conf. Series **865** (2017) 012001 doi:10.1088/1742-6596/865/1/012001

by Hassan and Rosen [10], which seems to be one of the most interesting generalizations of dRGT theory due to the change that the reference metric is introduced to be fully dynamical as the physical one. Note that the reference metric in the dRGT theory has been fixed as a non-dynamical metric. Along with the original (nonlinear) massive gravity, the massive bi-gravity has also been discussed extensively, see, for example, a recent interesting review paper [11] for more details.

Theoretically, inspired by the well-known higher dimensional theories, e.g., the string and super-gravity theories, one could think of higher dimensional scenarios of both massive gravity and bi-gravity. However, we have observed that many published papers on massive (bi-)gravity have focused only on four-dimensional spacetimes, in which only the first three graviton terms $\mathcal{L}_2$, $\mathcal{L}_3$, and $\mathcal{L}_4$ exist. Besides, there are a few papers discussing higher dimensional massive gravity and bi-gravity [12, 13]. However, they have not studied particular metrics such as the well-known Friedmann-Lemaitre-Robertson-Walker [14], Bianchi type I [15], and Schwarzschild-Tangherlini [16] metrics for higher dimensional scenarios of massive (bi-)gravity. Hence, we would like to investigate cosmological implications of both five-dimensional massive gravity and bi-gravity in [17, 18]. In particular, we are interested in examining whether the metrics mentioned above act as cosmological solutions of both five-dimensional massive gravity and bi-gravity. To do this, we first construct the graviton term $\mathcal{L}_5$, which vanishes in all four-dimensional spacetimes but does survive in any five- (or higher) dimensional one, by using the consequence of the Cayley-Hamilton theorem in algebra [19]. Then, we derive and solve constraints equations associated with the existence of reference metric to calculate the value(s) of massive graviton terms $\mathcal{L}_i$. Finally, we solve the corresponding field equations of physical metric to get the desired cosmological solutions.

The present article will be regarded as a summary of investigations on both five-dimensional massive gravity and bi-gravity. It will be organized as follows. A short introduction of our study has been written in the section 1. Basic setup and simple cosmological solutions of five-dimensional massive (bi-)gravity will be shown in the sections 2 and 3, respectively. Finally, conclusions will be given in the section 4.

2. Basic setup of five-dimensional massive (bi-)gravity

2.1. Cayley-Hamilton theorem and massive graviton terms

First, we would like to mention an action of the dRGT theory in four-dimensional spacetimes [7]:

$$S_{4d} = \frac{M_p^2}{2}\int d^4x\sqrt{-g}\Big\{R + m_g^2\left(\mathcal{L}_2 + \alpha_3\mathcal{L}_3 + \alpha_4\mathcal{L}_4\right)\Big\}, \tag{1}$$

where $m_g \neq 0$ the graviton mass, M_p the Planck mass, $\alpha_{3,4}$ free parameters, and $\mathcal{L}_i$ $(i = 2-4)$ the massive graviton terms (or interaction terms) given by

$$\mathcal{L}_2 = [\mathcal{K}]^2 - [\mathcal{K}^2], \tag{2}$$

$$\mathcal{L}_3 = \frac{1}{3}[\mathcal{K}]^3 - [\mathcal{K}][\mathcal{K}^2] + \frac{2}{3}[\mathcal{K}^3], \tag{3}$$

$$\mathcal{L}_4 = \frac{1}{12}[\mathcal{K}]^4 - \frac{1}{2}[\mathcal{K}]^2[\mathcal{K}^2] + \frac{1}{4}[\mathcal{K}^2]^2 + \frac{2}{3}[\mathcal{K}][\mathcal{K}^3] - \frac{1}{2}[\mathcal{K}^4]. \tag{4}$$

Here, square brackets used in the definition of $\mathcal{L}_i$ are understood as the trace of matrix [7], e.g.,

$$[\mathcal{K}] \equiv tr\mathcal{K}^\mu{}_\nu;\ [\mathcal{K}]^2 \equiv \left(tr\mathcal{K}^\mu{}_\nu\right)^2;\ [\mathcal{K}^2] \equiv tr\mathcal{K}^\mu{}_\alpha\mathcal{K}^\alpha{}_\nu, \tag{5}$$

with the definition of $\mathcal{K}^\mu{}_\nu$ is given by

$$\mathcal{K}^\mu{}_\nu \equiv \delta^\mu{}_\nu - \sqrt{g^{\mu\alpha}f_{ab}\partial_\alpha\phi^a\partial_\nu\phi^b}, \tag{6}$$

where $g_{\mu\nu}$ the physical metric, f_{ab} the reference (or fiducial) metric, and ϕ^a $(a = 0-3)$ the Stückelberg scalar fields added to give a manifestly diffeomorphism invariant description [6, 7]. It is noted that unlike the physical metric the fiducial metric in the dRGT massive gravity is non-dynamical. Therefore, its field equations will be algebraic rather than differential. For detailed investigations on physical and cosmological aspects of the dRGT theory, see an interesting review paper by de Rham [9].

However, if we introduce the fiducial metric to be fully dynamical as the physical metric, we will have the corresponding extension of the dRGT theory called a massive bi-gravity (or bi-metric gravity), whose action is defined as follows [10]

$$S_{4d,bi} = M_g^2 \int d^4x\sqrt{g}R(g) + M_f^2 \int d^4x\sqrt{f}R(f) + 2m^2 M_{eff}^2 \int d^4x\sqrt{g}\left(\mathcal{U}_2 + \alpha_3\mathcal{U}_3 + \alpha_4\mathcal{U}_4\right), \quad (7)$$

where $\mathcal{U}_i = \mathcal{L}_i/2$, while $R(g)$ and $R(f)$ denote the corresponding Ricci scalar of the physical metric $g_{\mu\nu}$ and reference metric $f_{\mu\nu}$, respectively. Additionally, M_{eff} is an effective Planck mass defined as follows [10]

$$M_{eff}^2 = \left(\frac{1}{M_g^2} + \frac{1}{M_f^2}\right)^{-1}, \quad (8)$$

with M_g and M_f are two other Planck masses of $g_{\mu\nu}$ and $f_{\mu\nu}$, respectively. It is also noted that $\mathcal{K}^\mu{}_\nu \equiv \delta^\mu{}_\nu - \sqrt{g^{\mu\sigma}f_{\sigma\nu}}$ in the context of bi-gravity theory. Of course, due to the dynamical property (or the existence of Ricci scalar $R(f)$), field equations of reference metric in bi-gravity will no longer be algebraic but differential as that of physical metric. For an interesting review of recent developments of bi-gravity theory, see [11].

As a result, we have been able to show that the four-dimensional graviton terms, $\mathcal{L}_2$, $\mathcal{L}_3$, and $\mathcal{L}_4$, can be reconstructed by applying the well-known Cayley-Hamilton theorem in linear algebra [19] for a 4×4 matrix $\mathcal{K}^\mu{}_\nu$. Mathematically, this theorem states that any $n\times n$ matrix K must obey its characteristic equation:

$$\mathcal{P}(K) \equiv K^n - \mathcal{D}_{n-1}K^{n-1} + \mathcal{D}_{n-2}K^{n-2} - \ldots + (-1)^{n-1}\mathcal{D}_1 K + (-1)^n \det(K) I_n = 0, \quad (9)$$

where $\mathcal{D}_{n-1} = trK \equiv [K]$ and $\mathcal{D}_{n-j}$ $(2 \le j \le n-1)$ are coefficients of the characteristic polynomial [19]. It is straightforward to show that the forms of the first three graviton terms, $\mathcal{L}_2$, $\mathcal{L}_3$, and $\mathcal{L}_4$ look similar to the determinants of $K_{2\times 2}$, $K_{3\times 3}$, and $K_{4\times 4}$, respectively. This is indeed a key observation to construct any higher dimensional graviton terms, $\mathcal{L}_{i>4}$ [11, 12, 13, 17]. As a result, it turns out that $\mathcal{L}_i = 2\det \mathcal{K}_{i\times i}$ with $i = 2-n$. For example, we have been able to build up five-, six-, and seven-dimensional graviton terms as specific demonstrations [17, 18]:

$$\frac{\mathcal{L}_5}{2} = \frac{1}{120}\Big\{[\mathcal{K}]^5 - 10[\mathcal{K}]^3[\mathcal{K}^2] + 20[\mathcal{K}]^2[\mathcal{K}^3] - 20[\mathcal{K}^2][\mathcal{K}^3] + 15[\mathcal{K}][\mathcal{K}^2]^2 - 30[\mathcal{K}][\mathcal{K}^4] + 24[\mathcal{K}^5]\Big\}, \quad (10)$$

$$\begin{aligned}\frac{\mathcal{L}_6}{2} &= \frac{1}{720}\Big\{[\mathcal{K}]^6 - 15[\mathcal{K}]^4[\mathcal{K}^2] + 40[\mathcal{K}]^3[\mathcal{K}^3] - 90[\mathcal{K}]^2[\mathcal{K}^4] + 45[\mathcal{K}]^2[\mathcal{K}^2]^2 - 15[\mathcal{K}^2]^3 + 40[\mathcal{K}^3]^2 \\ &\quad -120[\mathcal{K}^3][\mathcal{K}^2][\mathcal{K}] + 90[\mathcal{K}^4][\mathcal{K}^2] + 144[\mathcal{K}^5][\mathcal{K}] - 120[\mathcal{K}^6]\Big\}, \end{aligned} \quad (11)$$

$$\begin{aligned}\frac{\mathcal{L}_7}{2} &= \frac{1}{5040}\Big\{[\mathcal{K}]^7 - 21[\mathcal{K}]^5[\mathcal{K}^2] + 70[\mathcal{K}]^4[\mathcal{K}^3] - 210[\mathcal{K}]^3[\mathcal{K}^4] + 105[\mathcal{K}]^3[\mathcal{K}^2]^2 - 420[\mathcal{K}]^2[\mathcal{K}^2][\mathcal{K}^3] \\ &\quad +504[\mathcal{K}]^2[\mathcal{K}^5] - 105[\mathcal{K}^2]^3[\mathcal{K}] + 210[\mathcal{K}^2]^2[\mathcal{K}^3] - 504[\mathcal{K}^2][\mathcal{K}^5] + 280[\mathcal{K}^3]^2[\mathcal{K}] - 420[\mathcal{K}^3][\mathcal{K}^4] \\ &\quad +630[\mathcal{K}^4][\mathcal{K}^2][\mathcal{K}] - 840[\mathcal{K}^6][\mathcal{K}] + 720[\mathcal{K}^7]\Big\}. \end{aligned} \quad (12)$$

IOP Conf. Series: Journal of Physics: Conf. Series **865** (2017) 012001 doi:10.1088/1742-6596/865/1/012001

2.2. Five-dimensional massive gravity

For simplicity, we would like to investigate a five-dimensional massive gravity with the following action [17]:

$$S_{5d} = \frac{M_p^2}{2}\int d^5x\sqrt{-g}\Big\{R + m_g^2\left(\mathcal{L}_2 + \alpha_3\mathcal{L}_3 + \alpha_4\mathcal{L}_4 + \alpha_5\mathcal{L}_5\right)\Big\}, \tag{13}$$

here α_5 is an additional field parameter associated with the massive graviton term $\mathcal{L}_5$ defined in Eq. (10). As a result, the corresponding Einstein field equations for the physical metric turn out to be [12, 13, 17]

$$\left(R_{\mu\nu} - \frac{1}{2}Rg_{\mu\nu}\right) + m_g^2\left(X_{\mu\nu} + \sigma Y_{\mu\nu} + \alpha_5 W_{\mu\nu}\right) = 0, \tag{14}$$

where

$$X_{\mu\nu} = -\frac{1}{2}\left(\alpha\mathcal{L}_2 + \beta\mathcal{L}_3\right)g_{\mu\nu} + \tilde{X}_{\mu\nu}, \tag{15}$$

$$\tilde{X}_{\mu\nu} = \mathcal{K}_{\mu\nu} - [\mathcal{K}]g_{\mu\nu} - \alpha\left\{\mathcal{K}^2_{\mu\nu} - [\mathcal{K}]\mathcal{K}_{\mu\nu}\right\} + \beta\left\{\mathcal{K}^3_{\mu\nu} - [\mathcal{K}]\mathcal{K}^2_{\mu\nu} + \frac{\mathcal{L}_2}{2}\mathcal{K}_{\mu\nu}\right\}, \tag{16}$$

$$Y_{\mu\nu} = -\frac{\mathcal{L}_4}{2}g_{\mu\nu} + \tilde{Y}_{\mu\nu}, \tag{17}$$

$$\tilde{Y}_{\mu\nu} = \frac{\mathcal{L}_3}{2}\mathcal{K}_{\mu\nu} - \frac{\mathcal{L}_2}{2}\mathcal{K}^2_{\mu\nu} + [\mathcal{K}]\mathcal{K}^3_{\mu\nu} - \mathcal{K}^4_{\mu\nu}, \tag{18}$$

$$W_{\mu\nu} = -\frac{\mathcal{L}_5}{2}g_{\mu\nu} + \tilde{W}_{\mu\nu}, \tag{19}$$

$$\tilde{W}_{\mu\nu} = \frac{\mathcal{L}_4}{2}\mathcal{K}_{\mu\nu} - \frac{\mathcal{L}_3}{2}\mathcal{K}^2_{\mu\nu} + \frac{\mathcal{L}_2}{2}\mathcal{K}^3_{\mu\nu} - [\mathcal{K}]\mathcal{K}^4_{\mu\nu} + \mathcal{K}^5_{\mu\nu}, \tag{20}$$

with $\alpha = \alpha_3 + 1$, $\beta = \alpha_3 + \alpha_4$, and $\sigma = \alpha_4 + \alpha_5$ as additional parameters introduced for convenience. On the other hand, the corresponding constraint equations associated with the existence of reference metric turn out to be

$$t_{\mu\nu} \equiv \tilde{X}_{\mu\nu} + \sigma\tilde{Y}_{\mu\nu} + \alpha_5\tilde{W}_{\mu\nu} - \frac{1}{2}\left(\alpha_3\mathcal{L}_2 + \alpha_4\mathcal{L}_3 + \alpha_5\mathcal{L}_4\right)g_{\mu\nu} = 0. \tag{21}$$

According to a detailed investigation made in [20], the modified Einstein field equations (14) can be reduced to

$$\left(R_{\mu\nu} - \frac{1}{2}Rg_{\mu\nu}\right) - \frac{m_g^2}{2}\mathcal{L}_M g_{\mu\nu} = 0, \tag{22}$$

with $\mathcal{L}_M \equiv \mathcal{L}_2 + \alpha_3\mathcal{L}_3 + \alpha_4\mathcal{L}_4 + \alpha_5\mathcal{L}_5$ as the total massive graviton Lagrangian. The equation (22) implies an important result that the massive graviton terms act as an effective cosmological constant, $\Lambda_M \equiv -m_g^2\mathcal{L}_M/2$, as a consequence of the Bianchi identity, $\partial^\nu\mathcal{L}_M = 0$. It has been shown in [17] that this result turns out to be valid for a large class of physical and compatible reference metrics. It appears that once the constraint equations (21), which are indeed algebraic rather than differential, are solved, the corresponding values of Λ_M may be figured out [17].

2.3. Five-dimensional massive bi-gravity

Analogous to the five-dimensional massive gravity [17], the corresponding action of five-dimensional massive bi-gravity is defined to be [18]

$$S_{5d,bi} = M_g^2\int d^5x\sqrt{g}R(g) + M_f^2\int d^5x\sqrt{f}R(f)$$
$$+2m^2M_{eff}^2\int d^5x\sqrt{g}\Big(\mathcal{U}_2 + \alpha_3\mathcal{U}_3 + \alpha_4\mathcal{U}_4 + \alpha_5\mathcal{U}_5\Big), \tag{23}$$

with $\mathcal{U}_i = \mathcal{L}_i/2$ $(i = 2-5)$ and therefore $\mathcal{U}_M \equiv \mathcal{U}_2 + \alpha_3\mathcal{U}_3 + \alpha_4\mathcal{U}_4 + \alpha_5\mathcal{U}_5 = \mathcal{L}_M/2$. As a result, the corresponding five-dimensional Einstein field equations turn out to be [12, 13, 18]

$$M_g^2\left(R_{\mu\nu} - \frac{1}{2}Rg_{\mu\nu}\right) + m^2 M_{eff}^2 \mathcal{H}_{\mu\nu}^{(5)}(g) = 0, \tag{24}$$

where

$$\mathcal{H}_{\mu\nu}^{(5)}(g) \equiv X_{\mu\nu} + \sigma Y_{\mu\nu} + \alpha_5 W_{\mu\nu}. \tag{25}$$

Note that the tensors $X_{\mu\nu}$, $Y_{\mu\nu}$, and $W_{\mu\nu}$ have been defined in Eqs. (15), (17), and (19). In fact, Eq. (24) turns out to be identical to Eq. (14) if $M_f = 0$, or equivalently $M_{eff} = M_g$.

Now, for the reference metric, we can derive its field equations to be

$$\sqrt{f}M_f^2\left(R_{\mu\nu}(f) - \frac{1}{2}f_{\mu\nu}R(f)\right) + \sqrt{g}m^2 M_{eff}^2 s_{\mu\nu}^{(5)}(f) = 0, \tag{26}$$

where the tensor $s_{\mu\nu}^{(5)}(f)$ is given by

$$\begin{aligned} s_{\mu\nu}^{(5)}(f) \equiv\; & -\hat{\mathcal{K}}_{\mu\nu} + \left\{[\mathcal{K}] + \alpha_3\mathcal{U}_2 + \alpha_4\mathcal{U}_3 + \alpha_5\mathcal{U}_4\right\}f_{\mu\nu} + \alpha\left\{\hat{\mathcal{K}}_{\mu\nu}^2 - [\mathcal{K}]\hat{\mathcal{K}}_{\mu\nu}\right\} \\ & -\beta\left\{\hat{\mathcal{K}}_{\mu\nu}^3 - [\mathcal{K}]\hat{\mathcal{K}}_{\mu\nu}^2 + \mathcal{U}_2\hat{\mathcal{K}}_{\mu\nu}\right\} - \sigma\left\{\mathcal{U}_3\hat{\mathcal{K}}_{\mu\nu} - \mathcal{U}_2\hat{\mathcal{K}}_{\mu\nu}^2 + [\mathcal{K}]\hat{\mathcal{K}}_{\mu\nu}^3 - \hat{\mathcal{K}}_{\mu\nu}^4\right\} \\ & -\alpha_5\left\{\mathcal{U}_4\hat{\mathcal{K}}_{\mu\nu} - \mathcal{U}_3\hat{\mathcal{K}}_{\mu\nu}^2 + \mathcal{U}_2\hat{\mathcal{K}}_{\mu\nu}^3 - [\mathcal{K}]\hat{\mathcal{K}}_{\mu\nu}^4 + \hat{\mathcal{K}}_{\mu\nu}^5\right\}. \end{aligned} \tag{27}$$

where the hatted tensors are defined as $\hat{\mathcal{K}}_{\mu\nu} = \mathcal{K}_\mu^\sigma f_{\sigma\nu}$, $\hat{\mathcal{K}}_{\mu\nu}^2 = \mathcal{K}_\mu^\rho\mathcal{K}_\rho^\sigma f_{\sigma\nu}$, $\hat{\mathcal{K}}_{\mu\nu}^3 = \mathcal{K}_\mu^\rho\mathcal{K}_\rho^\delta\mathcal{K}_\delta^\sigma f_{\sigma\nu}$, $\hat{\mathcal{K}}_{\mu\nu}^4 = \mathcal{K}_\mu^\rho\mathcal{K}_\rho^\delta\mathcal{K}_\delta^\gamma\mathcal{K}_\gamma^\sigma f_{\sigma\nu}$, and $\hat{\mathcal{K}}_{\mu\nu}^5 = \mathcal{K}_\mu^\rho\mathcal{K}_\rho^\delta\mathcal{K}_\delta^\gamma\mathcal{K}_\gamma^\sigma\mathcal{K}_\sigma^\alpha f_{\alpha\nu}$ for convenience. It is clear that Eq. (26) will no longer be algebraic but differential. Hence, it is not easy to have the constant-like behavior of graviton terms in the context of bi-gravity. Note, however, that there are two constraint equations due to the Bianchi identities for the physical and reference metrics in the five-dimensional bi-gravity [12, 13, 18]

$$D_g^\mu \mathcal{H}_{\mu\nu}^{(5)}(g) = 0, \tag{28}$$

$$D_f^\mu\left[\frac{\sqrt{g}}{\sqrt{f}}s_{\mu\nu}^{(5)}(f)\right] = 0, \tag{29}$$

where we use notations D_g^μ and D_f^μ for the covariant derivatives in the physical and reference sectors, respectively. As a result, these two constraint equations (28) and (29) turn out be very useful in order to seek solutions because they lead the graviton terms to effective cosmological constants in the context of bi-gravity. It is worth noting that one of solutions making the graviton terms of four-dimensional bi-gravity constant is that the reference metrics are proportional to the physical ones [10, 11]. Hence, one might expect that this result will also hold in higher-dimensional scenarios of bi-gravity theory, especially the five-dimensional one.

3. Simple cosmological solutions

Armed with the field and constraint equations of both five-dimensional massive gravity and bi-gravity, we are now able to seek their physical and cosmological solutions. Indeed, the Friedmann-Lemaitre-Robertson-Walker, Bianchi type I, and Schwarzschild-Tangherlini metrics have been shown to be solutions of five-dimensional massive gravity and bi-gravity in [17] and [18], respectively. Now, we would like to remark some basic details of investigations carried out in these papers [17, 18].

3.1. Friedmann-Lemaitre-Robertson-Walker metrics

As a result, we have considered the Friedmann-Lemaitre-Robertson-Walker spacetime [14] for both physical and reference metrics of in both papers [17, 18]:

$$ds^2_{5d}(g) = -N_1^2(t)dt^2 + a_1^2(t)\left(d\vec{x}^2 + du^2\right), \tag{30}$$

$$ds^2_{5d}(f) = -N_2^2(t)dt^2 + a_2^2(t)\left(d\vec{x}^2 + du^2\right), \tag{31}$$

where u is the fifth dimension, while a_1 and a_2 are the scale factors of physical and reference metrics, respectively. Note that the unitary gauge for the Stückelberg fields, i.e., $\phi^a = x^a$, is normally used in the context of massive gravity [7, 9, 17], while the massive bi-gravity does not address the existence of Stückelberg fields [10, 11, 18].

As a result, we have been able to obtain the following $\mathcal{L}_M$ for the massive gravity as [17]

$$\begin{aligned}\mathcal{L}_M = {} & 2\Big[(\gamma\alpha_4 - \gamma_3)\Sigma^3 + 3(\gamma_2 - \gamma\gamma_3)\Sigma^2 + 3(\gamma\gamma_2 - \gamma_1)\Sigma - \gamma\gamma_1 + (3\gamma_1 - 3\gamma_2 + \gamma_3)\Big] \\ & + 2\Big\{\alpha_4\Sigma^3 - \alpha_5(\gamma - 1)(\Sigma - 1)^3 - 3[\gamma_3 - (\gamma - 1)\alpha_4]\Sigma^2 + 3[\gamma_2 - (\gamma - 1)(\gamma_3 + \alpha_4)]\Sigma \\ & + (\gamma - 1)(3\gamma_3 + 1) - \gamma_1\Big\}(\Sigma - 1). \end{aligned} \tag{32}$$

Note that additional parameters appearing in Eq. (32) are defined as

$$\begin{aligned} & [\mathcal{K}]^n = (5 - \gamma - 4\Sigma)^n; \ [\mathcal{K}^n] = (1 - \gamma)^n + 4(1 - \Sigma)^n; \ \gamma = \frac{N_2}{N_1}; \ \Sigma = \frac{a_2}{a_1}; \\ & \gamma_1 = 3 + 3\alpha_3 + \alpha_4; \ \gamma_2 = 1 + 2\alpha_3 + \alpha_4; \ \gamma_3 = \alpha_3 + \alpha_4. \end{aligned} \tag{33}$$

According to the method used in [20], we first solve the Euler-Lagrange equations of N_2 and a_2:

$$\frac{\partial \mathcal{L}_M}{\partial N_2} = 0; \ \frac{\partial \mathcal{L}_M}{\partial a_2} = 0, \tag{34}$$

which can be respectively reduced to the following equations:

$$\frac{\partial \mathcal{L}_M}{\partial \gamma} = 0; \ \frac{\partial \mathcal{L}_M}{\partial \Sigma} = 0. \tag{35}$$

Given the explicit definition of $\mathcal{L}_M$ in Eq. (32), the constraint equations (35) become

$$\alpha_5\hat{\Sigma}^3 + 4\alpha_4\hat{\Sigma}^2 + 6\alpha_3\hat{\Sigma} + 4 = 0, \tag{36}$$

$$\left(\alpha_5\hat{\Sigma}^3 + 3\alpha_4\hat{\Sigma}^2 + 3\alpha_3\hat{\Sigma} + 1\right)\hat{\gamma} + \left(\alpha_4\hat{\Sigma}^2 + 3\alpha_3\hat{\Sigma} + 3\right)\hat{\Sigma} = 0, \tag{37}$$

where $\hat{\Sigma} \equiv 1 - \Sigma$ and $\hat{\gamma} \equiv 1 - \gamma$ as additional variables. Of course, these equations can also be obtained from the constraint equations $t_{\mu\nu} = 0$ as shown in Eq. (21). It turns out that both Eqs. (36) and (37) are non-linear equations of $\hat{\Sigma}$. Hence, solving them will give us several values of $\hat{\Sigma}$ and therefore $\mathcal{L}_M$ (see [17] for more details). Once the value of $\mathcal{L}_M$ is figured out, the corresponding Einstein field equations (22) can be solved to give

$$a_1(t) = \exp\left[\sqrt{\frac{\Lambda_M}{6}}t\right], \tag{38}$$

where $\Lambda_M \equiv -m_g^2\mathcal{L}_M/2$. It is indeed the de Sitter expanding solution assuming that $\Lambda_M > 0$.

Now, we turn to discuss the massive bi-gravity, whose physical metric equations are given by [18]

$$6\tilde{M}_g^2 H_1^2 + \hat{\Sigma}\left(\sigma\hat{\Sigma}^3 + 4\beta\hat{\Sigma}^2 + 6\alpha\hat{\Sigma} + 4\right) = 0, \tag{39}$$

$$3\tilde{M}_g^2\left(\dot{H}_1 + 2H_1^2\right) + \hat{\gamma}\left(\sigma\hat{\Sigma}^3 + 3\beta\hat{\Sigma}^2 + 3\alpha\hat{\Sigma} + 1\right) + \hat{\Sigma}\left(\beta\hat{\Sigma}^2 + 3\alpha\hat{\Sigma} + 3\right) = 0, \tag{40}$$

along with its reference metric equations defined as

$$6\tilde{M}_f^2\left(1-\hat{\Sigma}\right)^4 H_2^2 - \hat{\Sigma}\left(1-\hat{\gamma}\right)^2\left(\alpha_5\hat{\Sigma}^3 + 4\alpha_4\hat{\Sigma}^2 + 6\alpha_3\hat{\Sigma} + 4\right) = 0, \tag{41}$$

$$3\tilde{M}_f^2\left(1-\hat{\Sigma}\right)^3\left(\dot{H}_2 + 2H_2^2 + \frac{\dot{\hat{\gamma}}}{1-\hat{\gamma}}H_2\right) - \left(1-\hat{\gamma}\right)\left[\left(\alpha_5\hat{\gamma}+\alpha_4\right)\hat{\Sigma}^3 + 3\left(\alpha_4\hat{\gamma}+\alpha_3\right)\hat{\Sigma}^2 + 3\left(\alpha_3\hat{\gamma}+1\right)\hat{\Sigma} + \hat{\gamma}\right] = 0, \tag{42}$$

where $H_i \equiv \dot{a}_i/a_i$ as the Hubble constants for the physical and reference metrics. In addition, we have set additional variables as $\tilde{M}_g^2 \equiv M_g^2/(m^2 M_{eff}^2)$ and $\tilde{M}_f^2 \equiv M_f^2/(m^2 M_{eff}^2)$, while $\hat{\gamma} \equiv 1 - N_2$ since N_1 has been set to be one. Besides these equations, we also have the other constraint equations coming from Eqs. (28) and (29):

$$\partial_0\left[\hat{\Sigma}\left(\sigma\hat{\Sigma}^3 + 4\beta\hat{\Sigma}^2 + 6\alpha\hat{\Sigma} + 4\right)\right] + 4H_1\left(\hat{\Sigma} - \hat{\gamma}\right)\left(\sigma\hat{\Sigma}^3 + 3\beta\hat{\Sigma}^2 + 3\alpha\hat{\Sigma} + 1\right) = 0, \tag{43}$$

$$\left(4H_1 - 4H_2 + \frac{3\dot{\hat{\gamma}}}{1-\hat{\gamma}} + \partial_0\right)\left(1-\hat{\gamma}\right)\hat{\Sigma}\left(\alpha_5\hat{\Sigma}^3 + 4\alpha_4\hat{\Sigma}^2 + 6\alpha_3\hat{\Sigma} + 4\right) + 4H_2\left(\hat{\Sigma} - \hat{\gamma}\right)\left(\sigma\hat{\Sigma}^3 + 3\beta\hat{\Sigma}^2 + 3\alpha\hat{\Sigma} + 1\right) = 0. \tag{44}$$

As discussed in [18], one of solutions to both equations (43) and (44) is that the reference metric is proportional to the physical one, i.e.,

$$f_{\mu\nu} = (1-\hat{C})^2 g_{\mu\nu}, \tag{45}$$

which is equivalent to $\hat{\gamma} = \hat{\Sigma} = \hat{C}$, where $\hat{C}$ is a constant obeying the following equation:

$$\sigma\hat{C}^5 - 2(\sigma - 2\beta)\hat{C}^4 + \left(\sigma - 8\beta + 6\alpha + \alpha_5\tilde{M}^2\right)\hat{C}^3 + 4\left(\beta - 3\alpha + \alpha_4\tilde{M}^2 + 1\right)\hat{C}^2 + 2\left(3\alpha + 3\alpha_3\tilde{M}^2 - 4\right)\hat{C} + 4\left(\tilde{M}^2 + 1\right) = 0, \tag{46}$$

with $\tilde{M}^2 \equiv \tilde{M}_g^2/\tilde{M}_f^2$ as a dimensionless parameter. It is noted that Eq. (46) is a result of an equality, $\hat{\Lambda}_0^g = (1-\hat{C})^2\hat{\Lambda}_0^f$, with $\hat{\Lambda}_0^g$ and $\hat{\Lambda}_0^f$ are effective cosmological constants of physical and reference sectors, whose values are defined as follows

$$\hat{\Lambda}_0^g = -\frac{\hat{C}}{\tilde{M}_g^2}\left(\sigma\hat{C}^3 + 4\beta\hat{C}^2 + 6\alpha\hat{C} + 4\right);\ \hat{\Lambda}_0^f = \frac{\hat{C}}{\tilde{M}_f^2(1-\hat{C})^4}\left(\alpha_5\hat{C}^3 + 4\alpha_4\hat{C}^2 + 6\alpha_3\hat{C} + 4\right). \tag{47}$$

More interestingly, this solution leads to the constant-like feature of the graviton terms $\mathcal{L}_M$, which will make the field equations in both physical and reference sectors simpler to solve. It is noted that Eq. (46) will reduce to

$$\alpha_5\hat{C}^3 + 4\alpha_4\hat{C}^2 + 6\alpha_3\hat{C} + 4 = 0, \tag{48}$$

in the massive gravity limit, in which $s^{(5)}_{\mu\nu} = 0$ since the reference metric $f_{\mu\nu}$ is non-dynamical. Of course, this equation is identical to Eq. (36) noting that $\hat{C} = \hat{\Sigma}$. As a result, solving the corresponding field equations of bi-gravity yields [18]

$$a_1(t) = \exp\left[\sqrt{\frac{\hat{\Lambda}_0^g}{6}}t\right]; \; a_2(t) = (1-\hat{C})\exp\left[\sqrt{\frac{\hat{\Lambda}_0^g}{6}}t\right]. \tag{49}$$

It is noted in the context of massive bi-gravity that $\hat{\Lambda}_0^g M_g^2$ is different from Λ_M derived from the massive graviton terms [18]

$$\begin{aligned}\Lambda_M &= -m^2 M_{eff}^2 \hat{C}\left[\left(\sigma\hat{C}^3 + 4\beta\hat{C}^2 + 6\alpha\hat{C} + 4\right) + \left(\hat{C}-1\right)\left(\alpha_5\hat{C}^3 + 4\alpha_4\hat{C}^2 + 6\alpha_3\hat{C} + 4\right)\right] \\ &= \hat{\Lambda}_0^g\left[M_g^2 + M_f^2(1-\hat{C})^3\right]. \end{aligned} \tag{50}$$

3.2. Bianchi type I metrics

This subsection is devoted to consider the five-dimensional Bianchi type I spacetime [15] for both physical and reference metrics of massive (bi-)gravity [17, 18]:

$$\begin{aligned} ds^2_{5d}(g) &= -N_1^2(t)dt^2 + \exp\left[2\alpha_1(t) - 4\sigma_1(t)\right]dx^2 + \exp\left[2\alpha_1(t) + 2\sigma_1(t)\right]\left(dy^2 + dz^2\right) \\ &\quad + \exp\left[2\beta_1(t)\right]du^2, \end{aligned} \tag{51}$$

$$\begin{aligned} ds^2_{5d}(f) &= -N_2^2(t)dt^2 + \exp\left[2\alpha_2(t) - 4\sigma_2(t)\right]dx^2 + \exp\left[2\alpha_2(t) + 2\sigma_2(t)\right]\left(dy^2 + dz^2\right) \\ &\quad + \exp\left[2\beta_2(t)\right]du^2. \end{aligned} \tag{52}$$

As a result, the corresponding total graviton term $\mathcal{L}_M$ turns out to be [17, 18]

$$\begin{aligned} \mathcal{L}_M &= 2\Big[\left(\gamma\alpha_4 - \gamma_3\right)AB^2 + \left(\gamma_2 - \gamma\gamma_3\right)B\left(2A+B\right) \\ &\quad + \left(\gamma\gamma_2 - \gamma_1\right)\left(A+2B\right) - \gamma\gamma_1 + \left(3\gamma_1 - 3\gamma_2 + \gamma_3\right)\Big] \\ &\quad + 2\Big\{\alpha_4 AB^2 - \alpha_5\left(\gamma-1\right)\left(A-1\right)\left(B-1\right)^2 - \left[\gamma_3 - \left(\gamma-1\right)\alpha_4\right]B\left(2A+B\right) \\ &\quad + \left[\gamma_2 - \left(\gamma-1\right)\left(\gamma_3+\alpha_4\right)\right]\left(A+2B\right) + \left(\gamma-1\right)\left(3\gamma_3+1\right) - \gamma_1\Big\}\left(C-1\right), \end{aligned} \tag{53}$$

where we have defined additional variables:

$$A = \epsilon\eta^{-2}; \; B = \epsilon\eta; \; C = \exp\left[\beta_2 - \beta_1\right]; \; \epsilon = \exp\left[\alpha_2 - \alpha_1\right]; \; \eta = \exp\left[\sigma_2 - \sigma_1\right]. \tag{54}$$

First, we discuss the massive gravity. Similar to the previous subsection for the FLRW metrics, we will first solve the Euler-Lagrange equations of scale factors of reference metric [17]:

$$\frac{\partial\mathcal{L}_M}{\partial N_2} = 0; \; \frac{\partial\mathcal{L}_M}{\partial\alpha_2} = 0; \; \frac{\partial\mathcal{L}_M}{\partial\sigma_2} = 0; \; \frac{\partial\mathcal{L}_M}{\partial\beta_2} = 0; \tag{55}$$

which are reduced to the following equations:

$$\frac{\partial\mathcal{L}_M}{\partial\gamma} = 0; \; \frac{\partial\mathcal{L}_M}{\partial A} = 0; \; \frac{\partial\mathcal{L}_M}{\partial B} = 0; \; \frac{\partial\mathcal{L}_M}{\partial C} = 0. \tag{56}$$

As a result, these equations can be expanded to be non-linear equations of A, B, C, and γ [17]

$$\begin{aligned} &\alpha_4 AB^2 - \gamma_3 B\left(2A+B\right) + \gamma_2\left(A+2B\right) - \gamma_1 + \Big[\alpha_4 B\left(2A+B\right) - \alpha_5\left(A-1\right)\left(B-1\right)^2 \\ &- \left(\gamma_3+\alpha_4\right)\left(A+2B\right) + 3\gamma_3 + 1\Big]\left(C-1\right) = 0, \end{aligned} \tag{57}$$

$$
\begin{aligned}
&(\gamma\alpha_4-\gamma_3)\,B^2+2\,(\gamma_2-\gamma\gamma_3)\,B-\gamma_1+\gamma\gamma_2\\
&+\Big\{\alpha_4B^2-\alpha_5\,(\gamma-1)\,(B-1)^2-2\,[\gamma_3-(\gamma-1)\,\alpha_4]\,B-(\gamma-1)\,(\gamma_3+\alpha_4)+\gamma_2\Big\}\,(C-1)=0,
\end{aligned}
\tag{58}
$$

$$
\begin{aligned}
&(\gamma\alpha_4-\gamma_3)\,AB+(\gamma_2-\gamma\gamma_3)\,(A+B)-\gamma_1+\gamma\gamma_2+\Big\{\alpha_4AB-\alpha_5\,(\gamma-1)\,(A-1)\,(B-1)\\
&-[\gamma_3-(\gamma-1)\,\alpha_4]\,(A+B)-(\gamma-1)\,(\gamma_3+\alpha_4)+\gamma_2\Big\}\,(C-1)=0,
\end{aligned}
\tag{59}
$$

$$
\begin{aligned}
&\alpha_4AB^2-\gamma_3B\,(2A+B)+\gamma_2\,(A+2B)-\gamma_1+\Big[\alpha_4B\,(2A+B)-\alpha_5\,(A-1)\,(B-1)^2\\
&-(\gamma_3+\alpha_4)\,(A+2B)+3\gamma_3+1\Big]\,(\gamma-1)=0.
\end{aligned}
\tag{60}
$$

As a result, solving these equations will yield several values of effective cosmological constant $\Lambda_M \equiv -m_g^2\mathcal{L}_M/2$, see [17] for a detailed investigation. Once the value of Λ_M is figured out, we will have the corresponding component equations of Einstein field equations:

$$
3\left(\dot\alpha_1^2-\dot\sigma_1^2+\dot\alpha_1\dot\beta_1\right)=\Lambda_M;\;\; 3\left(\ddot\alpha_1+2\dot\alpha_1^2+\dot\sigma_1^2\right)=\Lambda_M; \tag{61}
$$

$$
\ddot\sigma_1+\dot\sigma_1\left(3\dot\alpha_1+\dot\beta_1\right)=0;\;\; \ddot\beta_1-2\dot\alpha_1^2+2\dot\sigma_1^2+\dot\beta_1^2+\dot\alpha_1\dot\beta_1=0. \tag{62}
$$

It turns out that solutions to these field equations can be found to be [17]

$$
\begin{aligned}
\exp\left[3\alpha_1\right] &= \exp\left[3\alpha_0\right]\left[\cosh\left(3\tilde H_1 t\right)+\frac{\dot\alpha_0}{\tilde H_1}\sinh\left(3\tilde H_1 t\right)\right],\\
\exp\left[\beta_1\right] &= \exp\left[\beta_0\right]\left[\cosh\left(3\bar H_1 t\right)+\frac{\dot\beta_0}{3\bar H_1}\sinh\left(3\bar H_1 t\right)\right],\\
\sigma_1 &= \sigma_0+\sqrt{\dot\alpha_0^2+\dot\alpha_0\dot\beta_0-H_1^2}\int\Bigg\{\left[\cosh\left(3\tilde H_1 t\right)+\frac{\dot\alpha_0}{\tilde H_1}\sinh\left(3\tilde H_1 t\right)\right]\\
&\quad\times\left[\cosh\left(3\bar H_1 t\right)+\frac{\dot\beta_0}{3\bar H_1}\sinh\left(3\bar H_1 t\right)\right]\Bigg\}^{-1}dt,
\end{aligned}
\tag{63}
$$

where $\alpha_0\equiv\alpha_1(t=0)$, $\dot\alpha_0\equiv\dot\alpha_1(t=0)$, $\beta_0=\beta_1(t=0)$, $\dot\beta_0=\dot\beta_1(t=0)$, and $\sigma_0=\sigma_1(t=0)$ are initial values. In addition, some extra parameters existing in the above solutions can be understood as $\tilde H_1^2=4H_1^2/9(1-V_0)$, $\bar H_1^2=V_0\tilde H_1^2$, $H_1^2\equiv\Lambda_M/3$, and V_0 is a constant (see [17] for more details).

Now, we would like to consider the five-dimensional massive bi-gravity to see whether it admits the above Bianchi type I metric as its solution. First, we examine the following constraint equations coming from the Bianchi identities (28) and (29):

$$
\begin{aligned}
&g^{00}\partial_0\mathcal{H}_{00}^{(5)}-g^{11}\left[\Gamma_{11}^0(g)\mathcal{H}_{00}^{(5)}+\Gamma_{10}^1(g)\mathcal{H}_{11}^{(5)}\right]-2g^{22}\left[\Gamma_{22}^0(g)\mathcal{H}_{00}^{(5)}+\Gamma_{20}^2(g)\mathcal{H}_{22}^{(5)}\right]\\
&-g^{44}\left[\Gamma_{44}^0(g)\mathcal{H}_{00}^{(5)}+\Gamma_{40}^4(g)\mathcal{H}_{44}^{(5)}\right]=0
\end{aligned}
\tag{64}
$$

and

$$
\begin{aligned}
&f^{00}\left\{\partial_0\left[\frac{\sqrt{g}}{\sqrt{f}}s_{00}^{(5)}\right]-2\frac{\sqrt{g}}{\sqrt{f}}\Gamma_{00}^0(f)s_{00}^{(5)}\right\}-\frac{\sqrt{g}}{\sqrt{f}}f^{11}\left[\Gamma_{11}^0(f)s_{00}^{(5)}+\Gamma_{10}^1(f)s_{11}^{(5)}\right]\\
&-2\frac{\sqrt{g}}{\sqrt{f}}f^{22}\left[\Gamma_{22}^0(f)s_{00}^{(5)}+\Gamma_{20}^2(f)s_{22}^{(5)}\right]-\frac{\sqrt{g}}{\sqrt{f}}f^{44}\left[\Gamma_{44}^0(f)s_{00}^{(5)}+\Gamma_{40}^4(f)s_{44}^{(5)}\right]=0.
\end{aligned}
\tag{65}
$$

As a result, we see that one of possible solutions to satisfy these constraint equations is that the reference metric $f_{\mu\nu}$ is proportional to the physical metric $g_{\mu\nu}$, similar to the FLRW case [18]:

$$f_{\mu\nu} = (1-\tilde{C})^2 g_{\mu\nu}, \tag{66}$$

where $\tilde{C} = 1-\gamma = 1-A = 1-B = 1-C$ is a constant obeying the following equation:

$$\begin{aligned}&\sigma\tilde{C}^5 - 2(\sigma-2\beta)\tilde{C}^4 + \left(\sigma - 8\beta + 6\alpha + \alpha_5\tilde{M}^2\right)\tilde{C}^3 + 4\left(\beta - 3\alpha + \alpha_4\tilde{M}^2 + 1\right)\tilde{C}^2\\ &+2\left(3\alpha + 3\alpha_3\tilde{M}^2 - 4\right)\tilde{C} + 4\left(\tilde{M}^2+1\right) = 0.\end{aligned} \tag{67}$$

Furthermore, this solution also leads to the constant-like property of massive graviton term $\mathcal{L}_M$. Once the value of $\tilde{C}$ is solved, the corresponding value of effective cosmological constant $\Lambda_M \equiv -m_g^2\mathcal{L}_M/2$ can be determined. Thanks to this result, the following solutions can be defined to be [18]

$$\begin{aligned}\exp[3\alpha_1] &= \exp[3\alpha_0]\left[\cosh\left(3\tilde{H}_1 t\right) + \frac{\dot{\alpha}_0}{\tilde{H}_1}\sinh\left(3\tilde{H}_1 t\right)\right],\\ \exp[\beta_1] &= \exp[\beta_0]\left[\cosh\left(3\bar{H}_1 t\right) + \frac{\dot{\beta}_0}{3\bar{H}_1}\sinh\left(3\bar{H}_1 t\right)\right],\\ \sigma_1 &= \sigma_0 + \sqrt{\dot{\alpha}_0^2 + \dot{\alpha}_0\dot{\beta}_0 - \frac{\tilde{\Lambda}_0^g}{3}}\int\left\{\left[\cosh\left(3\tilde{H}_1 t\right) + \frac{\dot{\alpha}_0}{\tilde{H}_1}\sinh\left(3\tilde{H}_1 t\right)\right]\right.\\ &\left.\times\left[\cosh\left(3\bar{H}_1 t\right) + \frac{\dot{\beta}_0}{3\bar{H}_1}\sinh\left(3\bar{H}_1 t\right)\right]\right\}^{-1} dt,\end{aligned} \tag{68}$$

where $\tilde{H}_1^2 = 4\tilde{\Lambda}_0^g/27(1-V_0^g)$, $\bar{H}_1^2 = V_0^g\tilde{H}_1^2$, V_0^g is a constant, and $\tilde{\Lambda}_0^g$ is an effective cosmological constant of physical sector, whose definition is given by

$$\tilde{\Lambda}_0^g = -\frac{\tilde{C}}{\tilde{M}_g^2}\left(\sigma\tilde{C}^3 + 4\beta\tilde{C}^2 + 6\alpha\tilde{C} + 4\right). \tag{69}$$

It is noted that Λ_M existing in Eq. (63) is different from $\tilde{\Lambda}_0^g$ appearing in Eq. (68) due to the dynamical property of reference metric of bi-gravity. It is worth noting that the discussions on the stability of Bianchi type I solutions of both massive gravity and bi-gravity have been written in [17, 18].

3.3. Schwarzschild-Tangherlini metrics

In this third subsection, we will summarize main results investigated in [17, 18] for the Schwarzschild-Tangherlini metrics of the following form [16]:

$$ds_{5d}^2(g) = -N_1^2(r)\,dt^2 + \frac{dr^2}{F_1^2(r)} + \frac{r^2 d\Omega_3^2}{H_1^2(r)}, \tag{70}$$

$$ds_{5d}^2(f) = -N_2^2(r)\,dt^2 + \frac{dr^2}{F_2^2(r)} + \frac{r^2 d\Omega_3^2}{H_2^2(r)}, \tag{71}$$

with $d\Omega_3^2 = d\theta^2 + \sin^2\theta d\varphi^2 + \sin^2\theta\sin^2\varphi d\psi^2$. As a result, the corresponding $\mathcal{L}_M$ becomes

$$\begin{aligned}\mathcal{L}_M &= 2\left\{\left[\alpha_5\left(\mathcal{K}^2{}_2\right)^3 + 3\alpha_4\left(\mathcal{K}^2{}_2\right)^2 + 3\alpha_3\mathcal{K}^2{}_2 + 1\right]\mathcal{K}^0{}_0\mathcal{K}^1{}_1\right.\\ &\left.+\mathcal{K}^2{}_2\left[\alpha_4\left(\mathcal{K}^2{}_2\right)^2 + 3\alpha_3\mathcal{K}^2{}_2 + 3\right]\left(\mathcal{K}^0{}_0 + \mathcal{K}^1{}_1\right) + \left(\mathcal{K}^2{}_2\right)^2\left(\alpha_3\mathcal{K}^2{}_2 + 3\right)\right\},\end{aligned} \tag{72}$$

where
$$\mathcal{K}^0{}_0(r) = 1 - \frac{N_2}{N_1};\ \mathcal{K}^1{}_1(r) = 1 - \frac{F_1}{F_2};\ \mathcal{K}^2{}_2(r) = \mathcal{K}^3{}_3(r) = \mathcal{K}^4{}_4(r) = 1 - \frac{H_1}{H_2}. \quad (73)$$

Similar to the previous subsections, we first consider the massive gravity by deriving its constraint equations $\partial\mathcal{L}_M/\partial\mathcal{K}^0{}_0 = \partial\mathcal{L}_M/\partial\mathcal{K}^1{}_1 = \partial\mathcal{L}_M/\partial\mathcal{K}^2{}_2 = 0$ to be

$$\left(\alpha_5\mathcal{K}^1{}_1 + \alpha_4\right)\left(\mathcal{K}^2{}_2\right)^3 + 3\left(\alpha_4\mathcal{K}^1{}_1 + \alpha_3\right)\left(\mathcal{K}^2{}_2\right)^2 + 3\left(\alpha_3\mathcal{K}^1{}_1 + 1\right)\mathcal{K}^2{}_2 + \mathcal{K}^1{}_1 = 0, \quad (74)$$
$$\left(\alpha_5\mathcal{K}^0{}_0 + \alpha_4\right)\left(\mathcal{K}^2{}_2\right)^3 + 3\left(\alpha_4\mathcal{K}^0{}_0 + \alpha_3\right)\left(\mathcal{K}^2{}_2\right)^2 + 3\left(\alpha_3\mathcal{K}^0{}_0 + 1\right)\mathcal{K}^2{}_2 + \mathcal{K}^0{}_0 = 0, \quad (75)$$
$$\left[\alpha_5\left(\mathcal{K}^2{}_2\right)^2 + 2\alpha_4\mathcal{K}^2{}_2 + \alpha_3\right]\mathcal{K}^0{}_0\mathcal{K}^1{}_1 + \left[\alpha_4\left(\mathcal{K}^2{}_2\right)^2 + 2\alpha_3\mathcal{K}^2{}_2 + 1\right]\left(\mathcal{K}^0{}_0 + \mathcal{K}^1{}_1\right) + \left(\alpha_3\mathcal{K}^2{}_2 + 2\right)\mathcal{K}^2{}_2 = 0. \quad (76)$$

As a result, solving these non-linear equations of $\mathcal{K}^0{}_0$, $\mathcal{K}^1{}_1$, and $\mathcal{K}^2{}_2$ will imply several values of $\mathcal{L}_M$ (see [17] for more details). Thanks to the constant-like behavior of $\mathcal{L}_M$, the physical Einstein field equations will turn out to be of the usual form with an effective cosmological constant $\Lambda_M \equiv -m_g^2\mathcal{L}_M/2$. Indeed, the five-dimensional physical metric (70) can be solved to be [16, 17]

$$ds^2 = -f(r)dt^2 + \frac{dr^2}{f(r)} + r^2 d\Omega_3^2,$$
$$N_1^2(t,r) = F_1^2(t,r) = f(r) = 1 - \frac{\mu}{r^2} - \frac{\Lambda_M}{6}r^2,\ H_1^2(t,r) = 1, \quad (77)$$

where $\mu \equiv 8G_5M/(3\pi)$ the mass parameter, M the mass of source, and G_5 the 5-dimensional Newton constant. It appears that if Λ_M is positive or negative definite then the metric (77) will be called the Schwarzschild-Tangherlini-de Sitter or Schwarzschild-Tangherlini-anti-de Sitter black hole, respectively. Otherwise, we will have the pure Schwarzschild-Tangherlini black hole for $\Lambda_M = 0$.

For the five-dimensional massive bi-gravity, its Bianchi constraint equations are given by

$$g^{11}\left[\partial_r\mathcal{H}_{11}^{(5)} - 2\Gamma_{11}^1(g)\mathcal{H}_{11}^{(5)}\right] - g^{00}\left[\Gamma_{00}^1(g)\mathcal{H}_{11}^{(5)} + \Gamma_{01}^0(g)\mathcal{H}_{00}^{(5)}\right] - g^{22}\left[\Gamma_{22}^1(g)\mathcal{H}_{11}^{(5)} + \Gamma_{21}^2(g)\mathcal{H}_{22}^{(5)}\right]$$
$$-g^{33}\left[\Gamma_{33}^1(g)\mathcal{H}_{11}^{(5)} + \Gamma_{31}^3(g)\mathcal{H}_{33}^{(5)}\right] - g^{44}\left[\Gamma_{44}^1(g)\mathcal{H}_{11}^{(5)} + \Gamma_{41}^4(g)\mathcal{H}_{44}^{(5)}\right] = 0, \quad (78)$$
$$g^{33}\left[\Gamma_{33}^2(g)\mathcal{H}_{22}^{(5)} + \Gamma_{32}^3(g)\mathcal{H}_{33}^{(5)}\right] + g^{44}\left[\Gamma_{44}^2(g)\mathcal{H}_{22}^{(5)} + \Gamma_{42}^4(g)\mathcal{H}_{44}^{(5)}\right] = 0, \quad (79)$$
$$g^{44}\left[\Gamma_{44}^3(g)\mathcal{H}_{33}^{(5)} + \Gamma_{43}^4(g)\mathcal{H}_{44}^{(5)}\right] = 0, \quad (80)$$

along with

$$f^{11}\left\{\partial_r\left[\frac{\sqrt{g}}{\sqrt{f}}s_{11}^{(5)}\right] - 2\frac{\sqrt{g}}{\sqrt{f}}\Gamma_{11}^1(f)s_{11}^{(5)}\right\} - \frac{\sqrt{g}}{\sqrt{f}}f^{00}\left[\Gamma_{00}^1(f)s_{11}^{(5)} + \Gamma_{01}^0(f)s_{00}^{(5)}\right]$$
$$-\frac{\sqrt{g}}{\sqrt{f}}f^{22}\left[\Gamma_{22}^1(f)s_{11}^{(5)} + \Gamma_{21}^2(f)s_{22}^{(5)}\right] - \frac{\sqrt{g}}{\sqrt{f}}f^{33}\left[\Gamma_{33}^1(f)s_{11}^{(5)} + \Gamma_{31}^3(f)s_{33}^{(5)}\right]$$
$$-\frac{\sqrt{g}}{\sqrt{f}}f^{44}\left[\Gamma_{44}^1(f)s_{11}^{(5)} + \Gamma_{41}^4(f)s_{44}^{(5)}\right] = 0, \quad (81)$$
$$f^{33}\left[\Gamma_{33}^2(f)s_{22}^{(5)} + \Gamma_{32}^3(f)s_{33}^{(5)}\right] + f^{44}\left[\Gamma_{44}^2(f)s_{22}^{(5)} + \Gamma_{42}^4(f)s_{44}^{(5)}\right] = 0, \quad (82)$$
$$f^{44}\left[\Gamma_{44}^3(f)s_{33}^{(5)} + \Gamma_{43}^4(f)s_{44}^{(5)}\right] = 0. \quad (83)$$

IOP Conf. Series: Journal of Physics: Conf. Series **865** (2017) 012001 doi:10.1088/1742-6596/865/1/012001

As a result, we observe that one of simple solutions to the is that the physical and reference metrics are proportional to each other [18], i.e.,

$$f_{\mu\nu} = (1-\bar{C})^2 g_{\mu\nu}, \tag{84}$$

where $\bar{C}$ is a constant obeying the following equation:

$$\sigma\bar{C}^5 - 2(\sigma-2\beta)\bar{C}^4 + \left(\sigma - 8\beta + 6\alpha + \alpha_5\tilde{M}^2\right)\bar{C}^3 + 4\left(\beta - 3\alpha + \alpha_4\tilde{M}^2 + 1\right)\bar{C}^2$$
$$+2\left(3\alpha + 3\alpha_3\tilde{M}^2 - 4\right)\bar{C} + 4\left(\tilde{M}^2 + 1\right) = 0. \tag{85}$$

Similar to the massive gravity, solving the corresponding Einstein field equations of physical metric gives [16, 18]

$$g^{5d}_{\mu\nu}dx^\mu dx^\nu = -f(r)dt^2 + \frac{dr^2}{f(r)} + r^2 d\Omega_3^2,\; N_1^2(r) = F_1^2(r) = f(r) = 1 - \frac{\mu}{r^2} - \frac{\bar{\Lambda}_0^g}{6}r^2,$$
$$H_1^2(r) = 1,\; \bar{\Lambda}_0^g = -\frac{\bar{C}}{\tilde{M}_g^2}\left(\sigma\bar{C}^3 + 4\beta\bar{C}^2 + 6\alpha\bar{C} + 4\right). \tag{86}$$

4. Conclusions

We have mainly summarized our results on both five-dimensional massive gravity and bi-gravity, which have been published in two recent papers in [17, 18]. It turns out that both five-dimensional massive and bi-gravity are indeed physically non-trivial theories with a should-not-be-ignored $\mathcal{L}_5$ term, which disappears in all four-dimensional spacetimes but does exist in any five- (or higher) dimensional one. As a result, we have shown the useful method based on the Cayley-Hamilton theorem to construct arbitrary dimensional graviton terms $\mathcal{L}_i$ and therefore the five-dimensional graviton term $\mathcal{L}_5$. The following field equations of both five-dimensional physical and reference metrics have been derived consistently in order to discuss cosmological aspects. Note that the field equations of reference metric have been regarded as the constrained equations, which should be solved first to determine the corresponding values of massive graviton terms. It turns out that by counting the contribution of $\mathcal{L}_5$, the value(s) of effective cosmological constants coming from massive graviton terms $\mathcal{L}_M$ will be changed, and therefore the corresponding cosmological solutions will be different from that defined in scenarios without $\mathcal{L}_5$ [20]. To end this conclusion section, we would like to note again that $\mathcal{L}_5$ should not be ignored in all higher-than-four dimensional massive (bi-)gravity.

Acknowledgments

This research is supported in part by VNU University of Science, Vietnam National University, Hanoi.

References

[1] Fierz M and Pauli W 1939 On relativistic wave equations for particles of arbitrary spin in an electromagnetic field *Proc. R. Soc. Lond.* A **173** 211
[2] van Dam H and Veltman M J G 1970 Massive and massless Yang-Mills and gravitational fields *Nucl. Phys.* B **22** 397
[3] Zakharov V I 1970 Linearized gravitation theory and the graviton mass *JETP Lett.* **12** 312
[4] Vainshtein A I 1972 To the problem of nonvanishing gravitation mass *Phys. Lett.* B **39** 393
[5] Boulware D G and Deser S 1972 Can gravitation have a finite range? *Phys. Rev.* D **6** 3368
[6] Arkani-Hamed N, Georgi H and Schwartz M D 2003 Effective field theory for massive gravitons and gravity in theory space *Ann. Phys. (N.Y.)* **305** 96 (*Preprint* hep-th/0210184)
Dubovsky S L 2004 Phases of massive gravity *J. High Energy Phys.* JHEP10(2004)076 (*Preprint* hep-th/0409124)

Hinterbichler K 2012 Theoretical aspects of massive gravity *Rev. Mod. Phys.* **84** 671 (*Preprint* arXiv:1105.3735)

[7] de Rham C, Gabadadze G and Tolley A J 2011 Resummation of massive gravity *Phys. Rev. Lett.* **106** 231101 (*Preprint* arXiv:1011.1232)

de Rham C and Gabadadze G 2010 Generalization of the Fierz-Pauli action *Phys. Rev.* D **82** 044020 (*Preprint* arXiv:1007.0443)

[8] Hassan S F and Rosen R A 2012 Resolving the ghost problem in nonlinear massive gravity *Phys. Rev. Lett.* **108** 041101 (*Preprint* arXiv:1106.3344)

Hassan S F and Rosen R A 2012 Confirmation of the secondary constraint and absence of ghost in massive gravity and bimetric gravity *J. High Energy Phys.* JHEP04(2012)123 (*Preprint* arXiv:1111.2070)

Hassan S F, Rosen R A and Schmidt-May A 2012 Ghost-free massive gravity with a general reference metric *J. High Energy Phys.* JHEP02(2012)026 (*Preprint* arXiv:1109.3230)

Hassan S F, Schmidt-May A and von Strauss M 2012 Proof of consistency of nonlinear massive gravity in the Stuckelberg formulation *Phys. Lett.* B **715** 335 (*Preprint* arXiv:1203.5283)

Kluson J 2012 Remark about Hamiltonian formulation of nonlinear massive gravity in Stuckelberg formalism *Phys. Rev.* D **86** 124005 (*Preprint* arXiv:1202.5899)

Kluson J 2012 Nonlinear massive gravity with additional primary constraint and absence of ghosts *Phys. Rev.* D **86** 044024 (*Preprint* arXiv:1204.2957)

de Rham C, Gabadadze G and Tolley A J 2011 Helicity decomposition of ghost-free massive gravity *J. High Energy Phys.* JHEP11(2011)093 (*Preprint* arXiv:1108.4521)

de Rham C, Gabadadze G and Tolley A J 2012 Ghost free massive gravity in the Stuckelberg language *Phys. Lett.* B **711** 190 (*Preprint* arXiv:1107.3820)

Mirbabayi M 2012 A proof of ghost freedom in de Rham-Gabadadze-Tolley massive gravity *Phys. Rev.* D **86** 084006 (*Preprint* arXiv:1112.1435)

Golovnev A 2012 On the Hamiltonian analysis of non-linear massive gravity *Phys. Lett.* B **707** 404 (*Preprint* arXiv:1112.2134)

Kugo T and Ohta N 2014 Covariant approach to the no-ghost theorem in massive gravity *Prog. Theor. Exp. Phys.* 043B04 (*Preprint* arXiv:1401.3873)

[9] de Rham C 2014 Massive gravity *Living Rev. Relativity* **17** 7 (*Preprint* arXiv:1401.4173)

[10] Hassan S F and Rosen R A 2012 Bimetric gravity from ghost-free massive gravity *J. High Energy Phys.* JHEP02(2012)126 (*Preprint* arXiv:1109.3515)

Hassan S F and Rosen R A 2011 On non-linear actions for massive gravity *J. High Energy Phys.* JHEP07(2011)009 (*Preprint* arXiv:1103.6055)

[11] Schmidt-May A and von Strauss M 2016 Recent developments in bimetric theory *J. Phys.* A **49** 183001 (*Preprint* arXiv:1512.00021)

[12] Hassan S F, Schmidt-May A and von Strauss M 2013 Bimetric theory and partial masslessness with Lanczos-Lovelock terms in arbitrary dimensions *Classical Quantum Gravity* **30** 184010 (*Preprint* arXiv:1212.4525)

[13] Hinterbichler K and Rosen R A 2012 Interacting spin-2 fields *J. High Energy Phys.* JHEP07(2012)047 (*Preprint* arXiv:1203.5783)

Paulos M F and Tolley A J 2012 Massive gravity theories and limits of ghost-free bigravity models *J. High Energy Phys.* JHEP09(2012)002 (*Preprint* arXiv:1203.4268)

Huang Q G, Zhang K C and Zhou S Y 2013 Generalized massive gravity in arbitrary dimensions and its Hamiltonian formulation *J. Cosmol. Astropart. Phys.* JCAP08(2013)050 (*Preprint* arXiv:1306.4740)

[14] Nojiri S, Odintsov S D and Ogushi S 2001 Cosmological and black hole brane world universes in higher derivative gravity *Phys. Rev.* D **65** 023521 (*Preprint* hep-th/0108172)

Nojiri S, Odintsov S D and Ogushi S 2002 Friedmann-Robertson-Walker brane cosmological equations from the five-dimensional bulk (A)dS black hole *Int. J. Mod. Phys.* A **17** 4809 (*Preprint* hep-th/0205187)

[15] Maartens R, Sahni V and Saini T D 2001 Anisotropy dissipation in brane world inflation *Phys. Rev.* D **63** 063509 (*Preprint* gr-qc/0011105)

Campos A, Maartens R, Matravers D and Sopuerta C F 2003 Brane world cosmological models with anisotropy *Phys. Rev.* D **68** 103520 (*Preprint* hep-th/0308158)

[16] Tangherlini F R 1963 Schwarzschild field in n dimensions and the dimensionality of space problem *Nuovo Cim.* **27** 636

[17] Do T Q 2016 Higher dimensional nonlinear massive gravity *Phys. Rev.* D **93** 104003 (*Preprint* arXiv:1602.05672)

[18] Do T Q 2016 Higher dimensional massive bigravity *Phys. Rev.* D **94** 104022 (*Preprint* arXiv:1604.07568)

[19] Lipschutz S and Lipson M L 2009 *Schaum's Outline of Linear Algebra* (New York: McGraw-Hill) p 294

[20] Do T Q and Kao W F 2013 Anisotropically expanding universe in massive gravity *Phys. Rev.* D **88** 063006

Kao W F and Lin I-C 2014 Bianchi type I expanding universe in Weyl-invariant massive gravity *Phys. Rev.* D **90** 063003

IOP Conf. Series: Journal of Physics: Conf. Series **865** (2017) 012002 doi:10.1088/1742-6596/865/1/012002

Energy spectrum inverse problem of q-deformed harmonic oscillator and entanglement of composite bosons

Nguyen Anh Sang[1], Do Thi Thu Thuy[1], Nguyen Thi Ha Loan[1], Nguyen Tri Lan[2], and Nguyen Ai Viet[2]

[1]Hanoi Pedagogical University No.2, Nguyen Van Linh Street, Phuc Yen Town, Vinh Phuc, Vietnam.

[2]Institute of Physics, Vietnam Academy of Science and Technology, 10 Dao Tan, Thu Le, Ba Dinh, Ha Noi, Vietnam

E-mail: anhsangk13@gmail.com

Abstract. Using the simple deformed three-level model (D3L model) proposed in our early work, we study the entanglement problem of composite bosons. Consider three first energy levels are known, we can get two energy separations, and can define the level deformation parameter δ. Using connection between $q-$deformed harmonic oscillator and Morse-like anharmonic potential, the deform parameter q also can be derived explicitly. Like the Einstein's theory of special relativity, we introduce the observer effects: out side observer (looking from outside the studying system) and inside observer (looking inside the studying system). Corresponding to those observers, the outside entanglement entropy and inside entanglement entropy will be defined.. Like the case of Foucault pendulum in the problem of Earth rotation, our deformation energy level investigation might be useful in prediction the environment effect outside a confined box.

1. Introduction

Deformed Heisenberg algebras with q-deformed harmonic oscillator recently have been attracted a great attention and are a subject of intensive investigation. This approach is found some applications in various branches of physics and chemistry [1-6]. This method of q-deformed quantum mechanics was developed on the base of Heisenberg commutation relations, and the main parameter is the deformation parameter $q \in [0, 1]$.

In the atomic and molecular physics, the Morse potential plays an important role. This potential excellent describes the the interaction between atoms in diatomic and even polyatomic molecules. One-dimensional Morse-like effective potentials with three-parameter set have many application in condensed matter, bio-physics, nano science and quantum optics [7-10].

The relation between the problem of q-deformed harmonic oscillator and Morse potential was considered in [8]. The vibration energy levels of a Morse potential are rather good described by the energy levels of q-deformed harmonic oscillator.

The extended SU(2) model for Morse potential is also developed to compare with phenomenological Dunham's expansion and found a good agreement with experimental data for numerous diatomic molecules [8-10].

Considering deformed algebra is a mathematical object and atomic effective potential likes a physical model, the properties of q-deformed harmonic oscillator on the base of the Morse potential was investigated inversely way in [11, 12]. The infinite equal-step levels of non-deformed harmonic oscillator can be mapped to the energy spectrum of the motion in a parabolic potential. The finite unequal-step levels of q-deformed harmonic oscillator might be described as the motion in a Morse-like anharmonic potential.

In the our previous work [13], the simple deformed three-level model is proposed (D3L model). Consider a three-level system with the new parameter δ characterized the level deformation. The corresponding Morse-like potential can be constructed and then the deformed parameter q can be defined for this system by using the connection between q-deformed harmonic oscillator and Morse-like anharmonic potential.

In this work we continue investigate this D3L model in the frame work of composite boson approach developed in [14-15]. The deformation of energy levels can be characterized by the entanglement entropy between the constituents of composite bosons. Like Einstein's theory of special relativity, we introduce the observer effects: out side observer (looking from outside the studying system) and inside observer (looking inside the studying system). Corresponding to those observers, the outside entanglement entropy and inside entanglement entropy will be defined. Standard cases [14-15] are relating to the outside observer with outside entanglement entropy.

Very recently, gravitational waves from a Binary Black Hole Merger was observed [16] with the LIGO. In the past, the Foucault pendulum told us the rotation of the Earth. That are two examples what inside observers can got the information beyond their systems. That equivalent to the problem how the observer in a closed box know what happen outside the box?. We consider that our D3L model with the case of inside observer and inside entanglement entropy could be useful to this problem.

2. Inverse problem of deformed harmonic oscillator

Consider the first three energy levels E_0, E_1, $E_2$are known, we can define two parameters $\Delta_1 = E_1 - E_0$, and $\Delta_2 = E_2 - E_1$. Introduce the ratio parameter δ as the main parameter of our model [13]

$$\delta = \Delta E_2/\Delta E_1. \tag{1}$$

We have the deformation parameter q

$$q = \frac{1}{2-\delta} \longleftrightarrow \delta = 2 - \frac{1}{q}, \tag{2}$$

the relation between q and δ are presented in the Fig. 1a and Fig. 1b

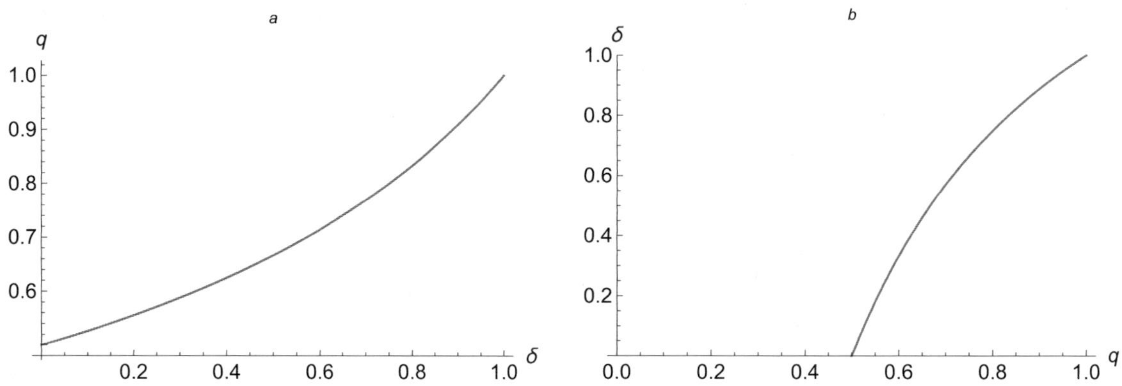

Figure 1. The relation between q and δ: a) q on δ, and b) δ on q.

In the connection between q-deformed harmonic oscillator and Morse-like anharmonic potential, the maximum level number n_{max} is an important parameter [13]. The value n_{max} is

$$n_{max} = \left[\frac{(2-\delta)}{2(1-\delta)} \right] = \left[\frac{1}{2(1-q)} \right], \tag{3}$$

where [...] is the integer part, and are presented in the Fig. 2 as a functions on δ (upper steps), and on q (lower steps).

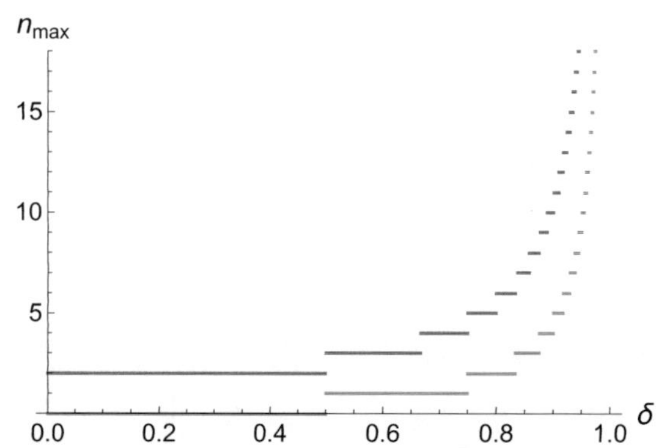

Figure 2. The relation value n_{max} on δ (upper steps), and on q (lower steps).

3. Entanglement entropy of composite bosons

Composite bosons or quasibosons are systems of quasi-particles, which consisted from two or more constituent particles. Among quasibosons in condensed mater and atomic physics there are excitons, atoms, molecules, polaritons, etc. In this part we focus on the case of two component composite bosons.

In order to find the energy dependence of the entanglement entropy we need the expression for the Hamiltonian H_C of the composite boson system with oscillation frequency ω_C. The Hamiltonian of deformed oscillators (deformed bosons) which provide realization of the composite bosons is [14-15]

$$H_C = \sum_n E_{Cn} = \sum_n \frac{1}{2} \hbar\omega_C \left[\varphi(n) + \varphi(n+1) \right]. \tag{4}$$

IOP Conf. Series: Journal of Physics: Conf. Series **865** (2017) 012002 doi:10.1088/1742-6596/865/1/012002

The structure function$\varphi(n)$ involving discrete deformation parameter f is quadratic in n, where n is the number of level energy or number of composite bosons

$$\varphi\left(n\right)=\left(1+\epsilon\frac{f}{2}\right)n-\epsilon\frac{f}{2}n^{2}, \tag{5}$$

here $\epsilon=+1$ for fermionic and $\epsilon=-1$ for bosonic constituents, and f is a real number.

Hamiltonian of the composite boson system can be written in the form

$$H_{C}=\sum_{n}E_{Cn}=\sum_{n}\hbar\omega_{C}\left\{\left(1+\epsilon\frac{f}{2}\right)\left(n+\frac{1}{2}\right)-\epsilon\frac{f}{2}\left(n+\frac{1}{2}\right)^{2}+\epsilon\frac{f}{8}\right\}. \tag{6}$$

Hamiltonian of the inverse problem with the Morse potential representation is

$$H_{I}=\sum_{n}E_{n}=\sum_{n}\Delta_{1}\left[(2-\delta)\left(n+\frac{1}{2}\right)-(1-\delta)\left(n+\frac{1}{2}\right)^{2}\right]. \tag{7}$$

The above two Hamiltonian are same form by replacing

$$\hbar\omega_{C}\rightarrow\Delta_{1},\ \epsilon\frac{f}{2}\rightarrow 1-\delta, \tag{8}$$

and neglecting some constant.

The entanglement entropy S_{ent} is defined as

$$S_{ent}=\ln\left(\frac{2}{f}\right). \tag{9}$$

Energy of one composite boson $n=1$ is

$$E=E_{C1}=\hbar\omega_{C}\left(\frac{3}{2}-\epsilon\frac{f}{2}\right) \tag{10}$$

Therefore the entanglement entropy $S_{ent}\left(E\right)$ equals

$$S_{ent}\left(E\right)=\ln\left[\frac{\epsilon}{\left(\frac{3}{2}-\frac{E}{\hbar\omega_{C}}\right)}\right], \tag{11}$$

or for the case of fermion constituents with energy in the interval $\hbar\omega_C/2\leq E\leq 3\hbar\omega_C/2$

$$S_{ent}\left(E\right)=-\ln\left(\frac{3}{2}-\frac{E}{\hbar\omega_{C}}\right), \tag{12}$$

and for the case of boson constituents with energy in the interval $3\hbar\omega_C/2\leq E\leq 5\hbar\omega_C/2$

$$S_{ent}\left(E\right)=-\ln\left(-\frac{3}{2}+\frac{E}{\hbar\omega_{C}}\right). \tag{13}$$

The entanglement entropy S_{ent} versus the value $\xi=E/\left(\hbar\omega_C\right)$ is presented in the Fig. 3 for two cases $\epsilon=+1$ with $1/2\leq\xi\leq 3\hbar\omega_C/2$ and $\epsilon=-1$ with $3/2\leq\xi\leq 5\hbar\omega_C/2$. In both cases, the entropy $S_{ent}\left(E\right)$ goes to infinity for the energy $E=3\hbar\omega_C/2$, which implies maximal entanglement between constituents.

IOP Conf. Series: Journal of Physics: Conf. Series **865** (2017) 012002 doi:10.1088/1742-6596/865/1/012002

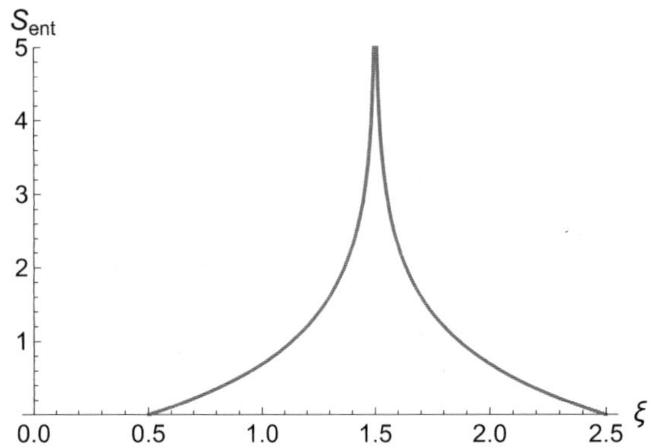

Figure 3. The entanglement entropy S_{ent} versus the value $\xi = E/(\hbar\omega_C)$ for two cases $\epsilon = +1$ with $1/2 \leq \xi \leq 3\hbar\omega_C/2$ and $\epsilon = -1$ with $3/2 \leq \xi \leq 5\hbar\omega_C/2$.

In this case the constituents (fermionic or bosonic) become most tightly bound within a quasiboson, and the quasiboson is most close to pure boson. On the contrary for $E = 3\hbar\omega_C/2$, $\epsilon = +1$, and $E = 5\hbar\omega_C/2$, $\epsilon = -1$, the entanglement entropy $S_{ent} = 0$ i.e. the constituents are unentangled. From physical viewpoint, in this case the constituents are in fact unbound.

4. Inverse problem of deformed harmonic oscillator and entanglement entropy

In order to applicable to the deformed harmonic oscillators we reformulate above problem of entanglement entropy of composite quasi-bosons following way:

(i) Instead of the fermionic or bosonic parameter of the constituents ϵ, we introduce the sign $\eta = \pm 1$ corresponds to the increasing or reducing cases of the energy gaps. Example: energy spectrum of Morse potential with reducing gaps is corresponding to the case $\eta = (+1)$, phonon spectrum in deformed universe model with increasing gaps is corresponding to the case $\eta = (-1)$.

(ii) Assume the observer effect like the Einstein theory of relativity. There are two cases: looking from outside and looking inside. a) Looking from out side case is similar the above case: one composite boson with two constituents. b) Looking from inside case: instead of the composite quasi-bosons with two fermionic or bosonic constituents we investigate one constituent boson entangling with another object.

4.1. Looking from outside case

Using the connection $\eta f/2 \leftrightarrow 1-\delta$ between the two problems, we can define outside entanglement entropy S_O in energy spectrum inverse problem as

$$S_O(\delta) = \ln\left[\frac{\eta}{(1-\delta)}\right], \tag{14}$$

where $0 \leq \delta \leq 0$ for the case of reducing gaps $\epsilon = +$, and $1 \leq \delta$ for the case of increasing gaps $\epsilon = +$.

Using the Morse potential representation of q-deformed harmonic oscillator, we have the connection $1-\delta \leftrightarrow (1/q) - 1$, the outside entanglement entropy S_I in this case can be expressed in the form

$$S_O(q) = \ln\left[\frac{\eta q}{(1-q)}\right]. \tag{15}$$

The values of outside entanglement entropy S_{O+} versus δ and q are plotted in the Fig. 4a and 4b, respectively.

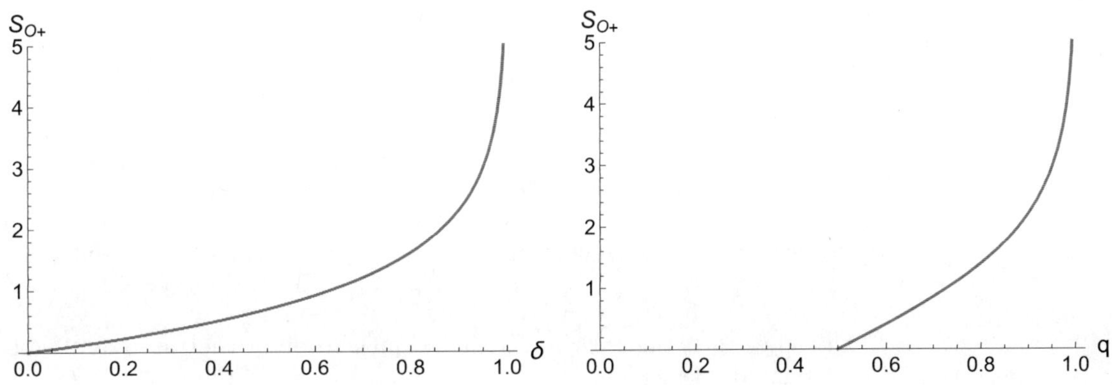

Figure 4. The values of outside entanglement entropy S_{O+} a) versus δ and b) versus q

We can see that outside entanglement entropy goes to infinity $S_{O+} \to \infty$ for $\delta \to 1$ or $q \to 1$ which implies maximal entanglement between two constituents. In this case the constituents (fermionic or bosonic) become tightly bound within a quasiboson, and the quasiboson is most close to pure boson. On the contrary, for $\delta \to 0$ or $q \to 1/2$ (only one energy level is existed), the outside entanglement entropy tents to zero $S_{O+} \to 0$) i.e. the constituents are unentangled. From physical viewpoint, in this case the constituents are in fact unbound.

4.2. Looking from inside case.
We consider a single non-interacting boson is non-deformed boson with equal energy levels $\delta = 1$ (presented by parabolic potential) and its inverse entanglement entropy $S_I = 0$. When a another object (second boson) is appeared near its location to form a composite boson, due to the interaction between the two constituent bosons, the energy spectrum of first boson is changed (presented by Morse potential). We can say the investigate boson recognizes the existence of the other object, this filling can be expressed by the inverse entanglement entropy (or inverse Shannon information)$S_I \neq 0$.

Using again the connection $\eta f/2 \leftrightarrow 1 - \delta$ between the two problems, and in analogy the Einstein theory of relativity we can define inside entanglement entropy S_I in energy spectrum inverse problem as

$$S_I(\delta) = \left(\ln \left[\frac{\eta}{(1-\delta)} \right] \right)^{-1}. \tag{16}$$

Using the Morse potential representation of q-deformed harmonic oscillator, we have the connection $1 - \delta \leftrightarrow (1/q) - 1$, the inside entanglement entropy S_I in this case can be expressed in the form

$$S_I(q) = \left(\ln \left[\frac{\eta q}{(1-q)} \right] \right)^{-1}. \tag{17}$$

The values of inside entanglement entropy S_{I+} versus δ and q are plotted in the figure 5, a) and b), respectively.

We can see that inside entanglement entropy goes to zero $S_I \to 0$ for $\delta \to 1$ or $q \to 1$ which implies minimal entanglement between the investigate boson and the object. In this case the

IOP Conf. Series: Journal of Physics: Conf. Series **865** (2017) 012002 doi:10.1088/1742-6596/865/1/012002

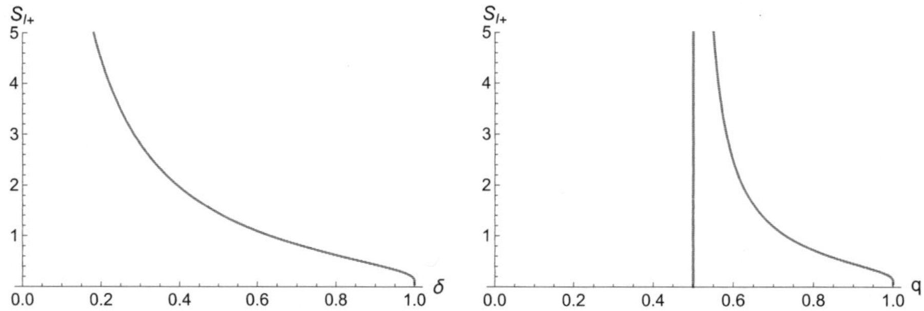

Figure 5. The values of inside entanglement entropy S_{I+} a) versus δ and b) versus q

investigated boson and the object become unbound within a quasiboson or they are not affected one to another. The investigated constituent quasiboson is most close to pure boson. On the contrary, for $\delta \to 0$ or $q \to 1/2$ (only one energy level is existed), the inside entanglement entropy tents to infinity $S_I \to \infty$) i.e. the investigate bonson and out box object are strongly entangled. From physical viewpoint, in this case the investigate boson and object are in tightly bound.

We can use the results of looking inside case to design a seismometer inside a closed box to detecting object outside the box by deformed energy levels.

5. Discussion

In this work using the our simple deformed three-level model (D3L model) proposed in [13], we study the entanglement problem of composite bosons.

Consider three first energy levels E_0, E_1, E_2 are known, we can get two energy separations $\Delta_1 = E_1 - E_0$, and $\Delta_2 = E_2 - E_1$ (consider $\Delta_2 \neq \Delta_1$ in general), and can define the level deformation parameter $\delta = (\Delta E_2/\Delta E_1)$. Using connection between q-deformed harmonic oscillator and Morse-like aharmonic potential, the deform parameter $q = 1/(2 - \delta)$ also can be derived explicitly. In the week-deform limit $\delta \to 1, q \to 1$, we back to the harmonic case with unique step levels $\Delta_2 = \Delta_1$. In the strong-deform limit $\delta \to 0$, $q \to 1/2$, we go to the two-level problem, where only the ground and first levels can be existed, and the other levels are collapsed, so $\Delta_2 = 0$.

The deformation of energy levels can be characterized by the entanglement entropy between the constituents of composite bosons. Like Einstein's theory of special relativity, we introduce the observer effects: out side observer O (looking from outside the studying system) and inside observer I (looking inside the studying system). Corresponding to those observers, the outside entanglement entropy S_O and inside entanglement entropy S_I will be defined.

Standard cases are relating to the outside observer with outside entanglement entropy. Outside entanglement entropy goes to infinity $S_{O+} \to \infty$ for $\delta \to 1$ or $q \to 1$ which implies maximal entanglement between two constituents. In this case the constituents (fermionic or bosonic) become tightly bound within a quasiboson, and the quasiboson is most close to pure boson. On the contrary, for $\delta \to 0$ or $q \to 1/2$ (only one energy level is existed), the outside entanglement entropy tents to zero $S_{O+} \to 0$) i.e. the constituents are unentangled. From physical viewpoint, in this case the constituents are in fact unbound.

The looking from inside case is interestingly. Inside entanglement entropy goes to zero $S_I \to 0$ for $\delta \to 1$ or $q \to 1$ which implies minimal entanglement between the investigate boson and the object. In this case the investigated boson and the object become unbound within a quasiboson or they are not affected one to another. The investigated constituent quasiboson is most close to pure boson. On the contrary, for $\delta \to 0$ or $q \to 1/2$ (only one energy level is existed), the inside entanglement entropy tents to infinity $S_I \to \infty$) i.e. the investigate boson and out box object are strongly entangled. From physical viewpoint, in this case the investigate boson and object are in tightly bound.

We can use the results of looking inside case to design a seismometer inside a closed box to detecting object outside the box by deformed energy levels. Like the case of Foucault pendulums in the problem of Earth rotation, and gravitation sensors LIGO, our deformation energy level investigation might be useful in prediction the environment effect outside a confined box effect (inside observer and outside effect) by using the inside entanglement entropy approach.

Acknowledgments

This research is funded by Vietnam National Foundation for Science and Technology Development (NAFOSTED) under grant number 103.01-2015.42.

References

[1] Mirzrahi S S, Camargo J P and Dodonov V V 2004 *J. Phys. A* **37** 3707
[2] Abe S 2009 *Phys. Rev. E* **79** 041116
[3] Lavagno A and Swamy P P 2010 *Physica A* **389** 993
[4] Algin A and Senay M 2012 *Phys. Rev. E* **85** 041123
[5] Andrade F M and Silva E O 2013 *Phys. Lett. B* **719** 467
[6] Tkachuck V M 2013 *Phys. Rev. A* **86** 062112
[7] Dong S H, Lemus R and Frank A 2002 *Int. J. Quant. Phys.* **86** 433
[8] Angelova M, Dobrev V and Franck A 2004 *Eu. Phys. Jour. D* **31** 27; Angelova A 2002 *Preprint* cond–mat/0201240
[9] Kaplan I G 2003 *Handbook of Molecular Physics and Quantum Chemistry* (Wiley)
[10] Lima E F and Hornos E M 2005 *J. Phys. B: At. Mol. Opt. Phys.* **38** 815
[11] Man Van Ngu, Ngo Gia Vinh, Nguyen Tri Lan, Luu Thi Kim Thanh and Nguyen Ai Viet 2015 *J. Phys.: Conf. Series* **627** 012021
[12] Ngo Gia Vinh, Man Van Ngu, Do Thi Nga, Nguyen Tri Lan, Luu Thi Kim Thanh and Nguyen Ai Viet 2015 *J. Phys.: Conf. Series* **627** 012022
[13] Nguyen Anh Sang, Do Thi Thu Thuy, Nguyen Thi Ha Loan, Nguyen Tri Lan and Nguyen Ai Viet 2016 *J. Phys.: Conf. Series* **726** 012019
[14] Gavrilik A M and Mishchenko Yu A 2012 *Phys. Lett. A* **376** 1596
[15] Gavrilik A M and Mishchenko Yu A 2012 *Preprint* quant-ph/1211.1907
[16] Venkateswara K, Hagedorn A C, Turner D M, Arp T and Gundlach H J 2014 *Rev. Sci. Instrum.* **85** 015005
[17] Abbott B P et al. 2016 *Phys. Rev. Let.* **116** 061102

IOP Conf. Series: Journal of Physics: Conf. Series **865** (2017) 012003 doi:10.1088/1742-6596/865/1/012003

Study of crystallization mechanisms of Fe nanoparticle

P H Kien[1], G T T Trang[2], P K Hung[2]

[1] Department of Physics, Thainguyen University of Education, 20 Luong Ngoc Quyen, Thainguyen, Vietnam.

[2] Department of Computational Physics, Hanoi University of Technology, 1 Dai Co Viet, Hanoi Vietnam.

E-mail: giapthuytrang@gmail.com

Abstract. In this paper, the nanoparticle (NP) Fe was investigated by means of molecular dynamics simulation. The crystallization mechanism was studied through the time evolution of crystal cluster and potential energies of different atom types. The simulation shows that the NP was crystallized into bcc crystal structure when it was annealed at 900 K for long times. At early stage of the annealing, small nuclei form in different places of NP and dissolve for short times. After long times some nuclei form and gather nearby which create the stable clusters in the core of NP. After that the crystal clusters grow in the direction to cover the core and then to spread into the surface of NP. Analyzing the energies of different type atoms, we found that the crystal growth is originated from specific atomic arrangement in the boundary region of crystal clusters.

1. Introduction

Recently, interest in NPs has increased strongly owing their unique properties [1, 2, 3, 4, 5]. For example, amorphous iron and iron-based NPs exhibits unique magnetic, optical and sorption properties, and this may lead to their advanced application in nanotechnology. Like bulk counterparts, NPs have different morphologies which are also of great interest for the physics community. NPs can be produced either in amorphous or crystalline states. The amorphous NP has two parts: the core which has the amorphous structure similar to the bulk counterpart, and the surface which has more porous structure. Clarifying in detail the local structure of NPs is important, because it is related to certain specific properties found for these materials. On other hands, amorphous NP is thermodynamically unstable and may be crystallized. The stability of amorphous NPs against crystallization plays an important role because of this related to their working ability in practice. The crystallization of amorphous NPs is studied mainly by experiments [6, 7, 8, 9, 10, 11, 12]. It was shown that the crystallization is realized via the nucleation mechanism. However, the crystallization mechanism at the atomic level is poorly understood. Especially, it is still unclear how the free surface affects on the nucleation and crystal growth. Crystallization mechanisms of materials can be studied by means of molecular dynamics (MD) simulation [13, 14, 15], since it is difficult to trace the behavior of atoms directly by experiments. In present work we conduct the MD simulation with using Pak-Doyama potentials to study the crystallization mechanisms of Fe NPs.

2. Calculation procedure

The MD simulation is performed for a system containing 10^4 Fe atoms with free boundary conditions and the Pak-Doyma potentials. Initially, all atoms are randomly placed in a sphere with radius of 34Å . Then the statistical relaxation is carried out until the system reached the equilibrium. This sample has heated to temperature of 300 K. The obtained sample then has relaxed isothermally (annealing) by 2.10^7 steps to prepare the amorphous NP. The well-equilibrated sample of amorphous NP is called 300-sample. We additionally have prepared two 300-samples to examine the influence of initial configurations. The result shows that no effect of initial configuration was found for the structural properties. In order to study the crystallization we have prepared second sample by heating the 300-sample to 900 K and then relaxing isothermally over 2.10^7 steps. This sample is called 900-sample. To analyze the structure of NP we determine the pair radial distribution function (PRDF) using the procedure reported in previous work [16].

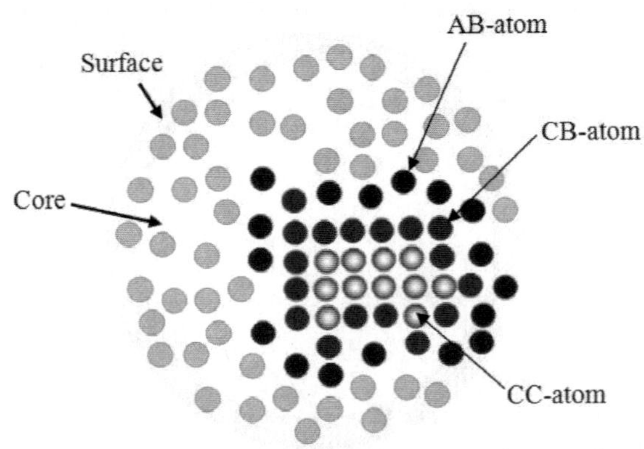

Figure 1. Schematic illustration of NP Fe.

In Figure 1 we schematically illustrate the NP. There is a crystal cluster in the sample. For the convenience the atoms belonging to amorphous and crystalline phases are called Am-atom and Cr-atom, respectively. In following we denote N_{Am} and N_{Cr} to the number of Am-atoms and Cr-atoms, respectively. There is a boundary region between amorphous and crystalline phases. We call Am-atoms, Cr-atoms in the boundary region AB-atom and CB-atom, respectively. The remaining Cr-atoms is called CC-atom. We denote N_{CB} and N_{CC} to the number of CB-atoms and CC-atoms, respectively. Obviously $N_{Cr}= N_{CB} + N_{CC}$. To determine a particular atom be Cr-atom or Am-atom, following criteria were applied. That is, the atom was identified as having the bcc configuration if it satisfies two conditions: (i) it has 14 neighbors; (ii) six among these neighbors have 4 neighbors and remaining ones have 6 neighbors in common with the given atom. The cutoff radius to determine the neighboring atom is equal to 3.35 Å. Such atom and its 14 neighbors belong to Cr-atoms and form a basic nucleus. An basic nucleus locating alone in the amorphous matrix comprises a CC-atom and 14 CB-atoms. Two basic nuclei are linking if they have at least an common Cr-atom. A crystal cluster contains a number of basic nuclei that each nucleus links at least to another nucleus.

3. Results and discussion

In Figure 2A we show the number of Cr-atoms, N_{Cr} as a function of steps. The process can be divided into three periods. In the first period we do not find any stable crystal cluster. There are only fewer nuclei which form and dissolve for short times. In the second period N_{Cr} significantly increases indicating the substantial growth of crystal clusters. In the third period, as expected, N_{Cr}fluctuates around a fix value which indicates that the crystal growth completes. The PRDFs determined at points in first and third periods are shown in Figure 2B. The PRDF found at third period has many peaks which demonstrate the crystalline structure. Another one determined at first period is similar to PRDF of the bulk sample which indicates that the structure of NP is still amorphous.

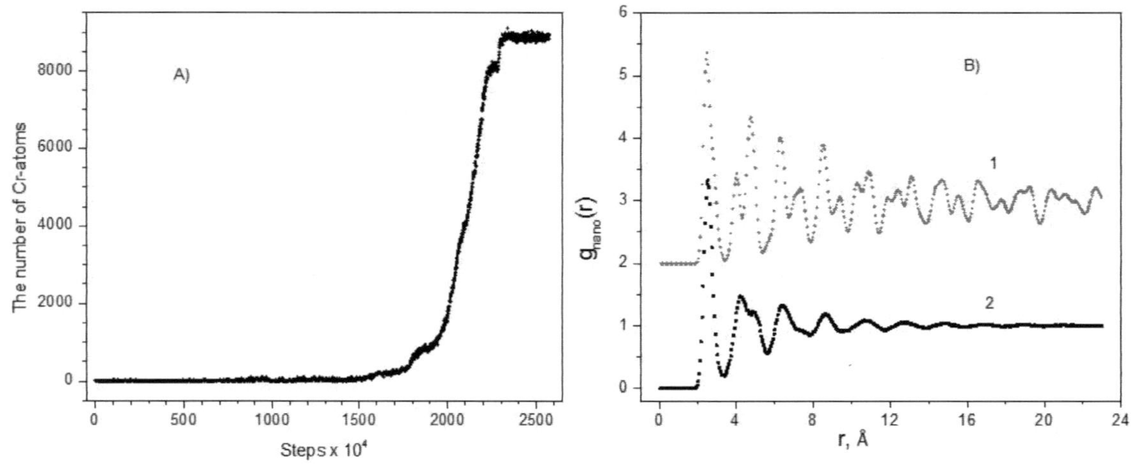

Figure 2. A) The dependence of number of Cr-atoms as a function of steps for 900-sample; B) the PRDFs determined at the moment in the first period (2) and in the third period (1).

Figure 3A shows the number N_{Cr} during 3.10^6 steps in the first period. One can see that N_{Cr} varies in the interval from 0 to 30 atoms. This means that the nuclei form and dissolve for short times. To identify what atoms are Cr-atoms during the considered time interval, at every MD steps we find Cr-atoms and record them. The number of Cr-atoms recorded is shown in Figure 3B. It can be seen that about 300 atoms take part in the formation of nuclei. It means that the amorphous NP contains a number of small groups of Am-atoms which gather nearby and locate at different places in NP. The Am-atoms of these groups arrange like the strongly distorted crystal lattice. Sometime they rearrange and form nuclei. Further examination shows that the number of recorded Cr-atoms in the core is much larger than in the surface. Obviously this is originated from the porous amorphous structure of the surface.

In Figure 4 we show the snapshots of arrangement of Cr-atoms which are generated at different moments in the second period. At early moment two crystal clusters were found in the core. Then these clusters grow and merge into a large cluster. After that the cluster expands over whole NP. It is worth to note that Cr-atoms gather nearby two clusters. From Figure 4 following remarks can be made. (i) Unlike first period, the stable crystal clusters have been created in the core; (ii) New nuclei now are not generated at different places, but only nearby the boundary regions of crystal clusters; (iii) The crystal growth is inhibited when the boundary region of crystal clusters moves to the surface. As such, the frequency of nuclei formation in the boundary regions is much bigger than one in the amorphous phase. The crystal cluster grows in the direction to cover the core and then to spread into the surface. The fully crystallized NP consists of the core which has the crystalline bcc structure and the surface which has the porous structure.

Consider a small crystal cluster. The CC-atoms and CB-atoms can rearrange so that the crystal cluster dissolves. This rearrangement depends on the mean potential energy per atom (MEPA) for CB-atoms, CC-atoms and AB-atoms. In particular, if MEPA of CB-atoms and CC-atoms is close to MEPA of AB-atoms, then such rearrangement is easy realized. In the case when

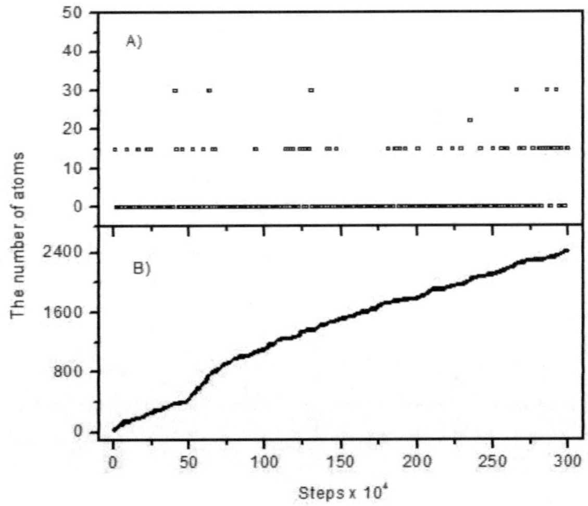

Figure 3. A) The number of Cr-atoms detected during 3.10^6 steps at the first period; B) The number of Cr-atoms recorded within this time interval.

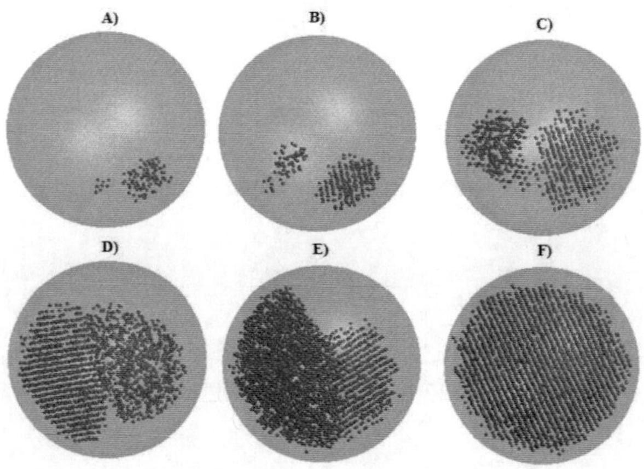

Figure 4. Snapshot of Cr-atom arrangement: A) N_{Cr} = 188; B) N_{Cr}=568; C) N_{Cr}=1651; D) N_{Cr}=4440; E) N_{Cr}=6162; F) N_{Cr}=8907.

the MEPA of those atoms is significantly smaller than one of AB-atoms, such rearrangement in contrast is less probable. Therefore, useful information on how the nuclei form, can be derived from MEPA for CB-atoms, CC-atoms and AB-atoms.

The time dependence of MEPA for different clusters is shown in Figure 5. One can see that for small crystal clusters, MEPA of CB-atoms is close to one of AB-atoms, and MEPA of CC-atoms fluctuates largely. It means that the crystal clusters are unstable and the distortion degree varies largely for different crystal clusters. As the size of the crystal clusters increases, MEPA of AB-atoms increases which is related to that the crystal cluster spreads into the surface where the energy of atoms is bigger than ones in the core. In the case of large crystal clusters, MEPA of different type atoms decreases in following order: AB-atom $\rightarrow$ CB-atom $\rightarrow$ CC-atom. It follows that the growing of the cluster leads to decreasing the energy of NP, while the shrinking of the cluster to increasing the energy of NP.

The MEPA of different type atoms as a function of number of Cr-atoms in the cluster is

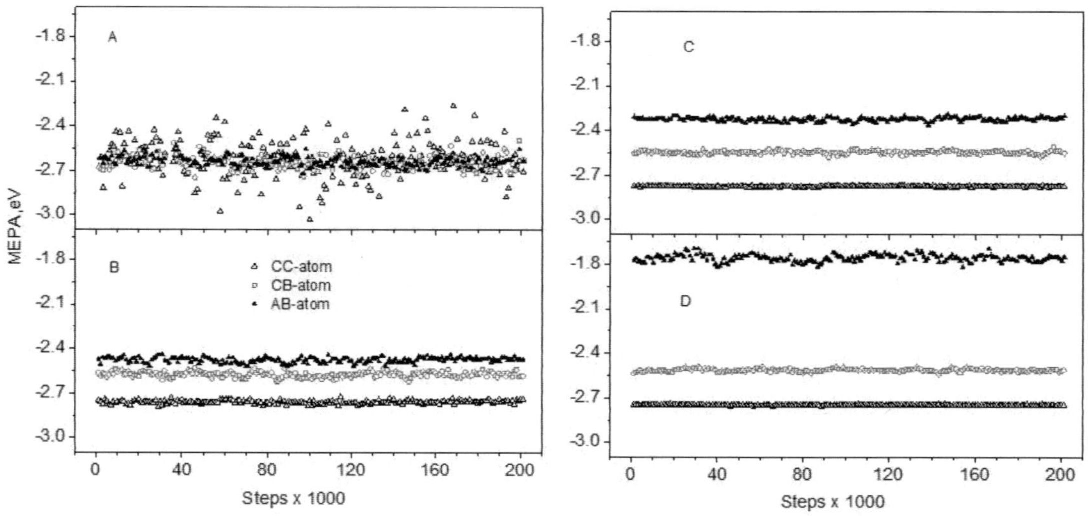

Figure 5. The MEPA of different type atoms as a function of steps for the crystal clusters: A) N_{Cr}=15-85 atoms; B) N_{Cr} =600-700 atoms; C) N_{Cr} =4100-4400 atoms; D) N_{Cr} =8900-9100 atoms.

shown in Figure 6. There is a point of 120 atoms where the MEPA of CC-atoms and CB-atoms is smaller than one of AB-atoms. This point corresponds to the so called critical size. The clusters with the size bigger than critical value are stable and tend to grow; otherwise they are unstable.

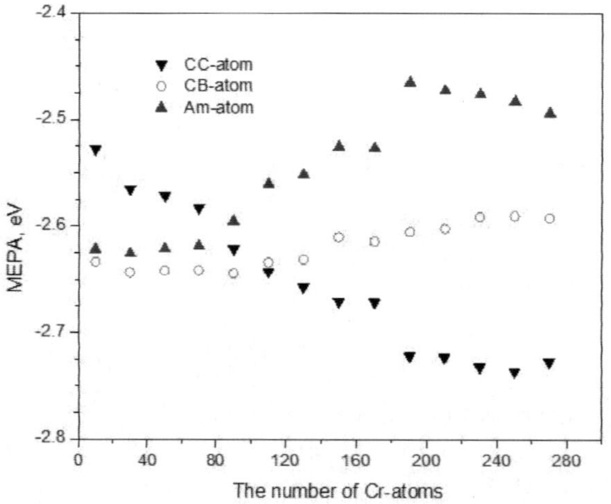

Figure 6. The dependence of MEPA as a function of the number of Cr-atoms in the crystal cluster.

In summary it is shown that at early stage of annealing the small nuclei form in different places of NP. Due to porous structure of the surface, the frequency of nuclei formation in the core is much larger than in the surface. MEPA of CB-atoms and CC-atoms of small clusters is close to the MEPA of AB-atoms. The nuclei are unstable and dissolve for short times. After long annealing a number of nuclei form nearby, which create a stable crystal cluster in the core of

NP. Unlike small cluster, MEPA of different type atoms for stable clusters decreases in following order: AB-atom → CB-atom → CC-atom. Such order leads to that the crystal cluster prefers to growing rather than to shrinking. The crystal growth is inhibited when the boundary region of crystal cluster moves to the surface. Hence, the crystal cluster grows in the direction to cover the core and then to spread into the surface. The fully crystallized sample consists of the crystalline phase in the core and amorphous phase in the surface.

4. Conclusion

In this paper, the crystallization of Fe NPs has been investigated by means of molecular dynamics simulation. The nucleation mechanism is studied through the time evolution of crystal clusters and the potential energy of different type atoms. According to our simulation, the crystallization occurred when the amorphous sample was annealed for long times at 900 K. At early stage of the annealing the small crystal clusters form in different places of NP and dissolve for short times. After long annealing, the stable crystal clusters have been created in the core of NP. After that these clusters grow in the direction to cover the core and then to spread into the surface. The fully crystallized NP consists of the core with bcc crystal structure and the surface with porous amorphous structure. Analyzing the potential energies of different type atoms, we found that unlike small cluster, for large cluster MEPA of CC-atoms, CB-atoms and AB-atoms decreases in following order: AB-atom → CB-atom → CC-atom. It means that the crystal growth is originated from specific atomic arrangement in the boundary region between amorphous and crystalline phases.

Acknowledgement

This research is funded by Vietnam National Foundation for Science and Technology Development (NAFOSTED) under grant number 103.05-2015.16.

References

[1] Yasushi S, Yusuke W and Toshio S 2009 *Chemical Physics Letters* **475** 264-268
[2] Levchenko E V, Evteev A V, Riley D P, Belova I V and Murch G E 2010 *Computational Materials Science* **47** 712-720
[3] Hoang V V and Ganguli D 2012*Physics Reports* **518** 81-140
[4] Carroll K J, Pitts J A, Zhang K, Pradhan A K and Carpenter E E 2010 *J. Appl. Phys.* **107** 09A302
[5] Qin S, Lei W, Liu D, Lamb P and Chen Y 2013 *Materials Letters* **91** 5-8
[6] Changsheng X, Junhui H, Run W and Hui X 1999 *NanoStructured Mater* **11** 1061-1066
[7] Alcoutlabi M and McKenna G B 2005 *J. Phys. Condens. Matter* **17** R461
[8] Gao L and Zhang Q 2001 *Scr. Mater.* **44** 1195-1198
[9] Zhang H and Banfield J F 2002 *J. Chem. Mater.* **14** 4145-4154
[10] Madras G and McCoy B J 2007 *Cryst. Growth and Design* **7** 250-253
[11] Pan C, Shen P and Chen S Y 2007 *J. Cryst. Growth* **299** 393-398
[12] Epifani M, Pellicer E, Arbiol J, Sergent N, Pargnier T and Morante J R 2008 *Langmuir* **24** 11182-11188
[13] Hou Z Y, Liu L X and Liu R S 2009 *Modelling Simul. Mater. Sci. Eng.* **17** 035001
[14] Hyeok S J, Cheol L S, Joo L B and Yoo S J 2003 *Journal of Crystal Growth* **250** 558-564
[15] Zhang Y, Wang L and Wang W 2007 *J. Phys. Condens. Matter* **19** 196106
[16] Kien P H, Lan M T, Dung N T and Hung P K 2014 *Int. J. Mod. Phys. B* **28** 1450155

IOP Conf. Series: Journal of Physics: Conf. Series **865** (2017) 012004 doi:10.1088/1742-6596/865/1/012004

The main scattering mechanisms in Single-side modulation doped square quantum wells

Tran Thi Hai, Nguyen Thi Dung, Nguyen Manh An
Faculty of Natural Science, Hong Duc University, Thanh Hoa City, Vietnam
E-mail: tranthihai@hdu.edu.vn

Abstract. We present a theory of the transport of carrier confined in the conduction band of a single-side modulation doped square quantum wells. Beside the well-known the scattering mechanisms such as surface roughness (SR), remote impurities (RI), and misfit deformation potential (DP), the theory induces acoustic phonon due to the deformation potential and acoustic phonon due to the piezoelectric coupling. In the result, we prove that acoustic phonon scattering has revealed very large mobility values. We have also consider acoustic-phonon partial mobility dependence on temperature for single-side modulation doped square quantum wells.

1. Introduction

The calculation of the carrier mobility in semiconductor heterostructures has received great attention in recent year due to their importance in device applications. In order to upgrade the electron mobility of the above-quoted QWs, one needs to identify the key scattering mechanisms limiting the transport properties of their two-dimensional electron and holes gas (2DEG and 2DHG), and reduce their detrimental effects. It is well known [1, 2] that the best way for this purpose is to study the dependence of 2DEG and 2DHG mobility on experimental conditions such as sample temperature, carrier density, and channel width.

It is worth mentioning that although the above dependencies were explored by a number of authors, they still remain as challenging problems. Firstly, the key scattering mechanisms for these QWs are a subject under debate. The previous interpretations of some experimental findings are quite different even due to one and the same research group [7, 8]. From the two-dimensional holes gas mobility dependence on temperature ($100K$) and on channel width some authors [3, 4, 8] assumed surface roughness scattering to be the key mechanism. However, from the carrier density dependence of the mobility and the transport to quantum lifetime ratio, the others [8] assumed ionized impurity scattering to be dominant. Secondly, in their calculations the roughness-related misfit deformation potential scattering has been ignored, which was proven to be important for strained Si [16] and SiGe [19] channels.

Just recently, some theoretical [18, 20] studies have indicated the well-known the scattering mechanisms such as surface roughness, misfit deformation potential and remote impurity.. To date, the theory of the key scattering mechanisms has been developed for a single-side modulation doped square quantum wells [20], however, there is not theory for mobility dependence on temperature. Thus, the goal of this paper is to provide a theory of the main scattering mechanisms in Single-side modulation doped square quantum wells, we develop a variational approach to the description of quantum confinement in bent-band infinite square QWs. For strained GaAs channel in GaAs/ AlGaAs single asymmetric quantum well, we incorporate all possible scattering mechanisms, especially optical phonons (dominant at high temperatures).

2. Theoretical determinations

As a prototype, we are dealing with a heterostructure made of cubic crystals, which is composed of a strained well layer grown pseudomorphically along the axis between two barrier ones. The well layer forms a conduction channel of width L in the region $|z| < L/2$. The sample is modulation doped in a region of thickness L_d in the surface-side barrier $z < -L/2$ and separated from the channel by a spacer layer of thickness L_s. The confinement may be plausibly described by an asymmetric wave function as follows:[21]

$$\zeta(z) = \begin{cases} B\sqrt{\pi/L}\cos(\pi z/L)e^{-cz/L} & for\ \ |z| \leq L/2, \\ 0 & for\ |z| > L/2, \end{cases} \tag{1}$$

with L as the well width. Here, B and c are parameters to be determined. The former may be given in terms of the latter via the normalization of the wave function. Thus, there is a single independent parameter, say c, which is, following Eq. (1), regarded as a measure of the band-bending effect on the carrier distribution. The band-bending parameter c is determined from the requirement that the wave function of the ground-state subband is to minimize the total energy per particle. In the bent-band model, besides the barrier potentials this energy includes the Hartree potential created by ionized impurities and charge carriers. In Ref. [21] we achieved an analytic expression for this energy, which enables a tractable variational evaluation of c.

The particles moving along the in-plane are scattered by various disorder sources, which are characterized by some random fields. Scattering by a Gaussian random field is specified by its autocorrelation function in wave vector space [15]. Here $U(\mathbf{q})$ is a two-dimensional Fourier transform of the unscreened scattering potential weighted with an envelope wave function:

$$U(\mathbf{q}) = \int_{-\infty}^{+\infty} dz|\zeta(z)|^2 U(\mathbf{q}, z). \tag{2}$$

The transport (τ_t) and quantum (τ_q) lifetimes are represented in terms of the ACF for each disorder by [12, 14]

$$\frac{1}{\tau_t} = \frac{1}{(2\pi)^2\hbar E_F}\int_0^{2k_F} dq \int_0^{2\pi} d\varphi \frac{q^2}{(4k_F^2 - q^2)^{1/2}} \frac{\langle|U(\mathbf{q})|^2\rangle}{\varepsilon^2(q)}, \tag{3}$$

and

$$\frac{1}{\tau_q} = \frac{1}{(2\pi)^2\hbar E_F}\int_0^{2k_F} dq \int_0^{2\pi} d\varphi \frac{2k_F^2}{(4k_F^2 - q^2)^{1/2}} \frac{\langle|U(\mathbf{q})|^2\rangle}{\varepsilon^2(q)}. \tag{4}$$

Here, $\mathbf{q} = (q, \varphi)$ is the 2D momentum transfer due to a scattering event in the in-plane (in polar coordinates), $q = |\mathbf{q}| = 2k_F\sin(\vartheta/2)$ with ϑ as a scattering angle. The Fermi energy is given by $E_F = \hbar^2k_F^2/2m^*$, with k_F as the Fermi wave number fixed by the sheet hole density p_s: $k_F = \sqrt{2\pi p_s}$.

The dielectric function $\varepsilon(q)$ in Eqs. (3) and (4) takes account of the screening of scattering potentials by the 2D carriers. As usual, this is evaluated within the random phase approximation, [15]

$$\varepsilon(q) = 1 + \frac{q_s}{q}F_S(q)\ [1 - G(q)],\ \ for\ \ q \leq 2k_F. \tag{5}$$

Here, $q_s = 2m^*e^2/\varepsilon_L\hbar^2$ is the inverse 2D Thomas-Fermi screening length, with ε_L as the dielectric constant of the well layer. The local field corrections due to a many-body exchange effect in the in-plane are quantified by $G(q) = q/2\sqrt{q^2 + k_F^2}$. The screening form factor $F_S(q)$ takes account of the extension of particle states along the growth direction. With the wave function from Eq. (1), $F_S(q)$ is analytically expressed in terms of the band-bending parameter c.

The charge carriers are in general expected to experience the following scattering mechanisms: (i) acoustic-phonon, (ii) remote impurities (RI), (iii) surface roughness (SR), and (iv) misfit deformation potential (DP) therefrom. The overall lifetime is determined by the ones for individual disorders according to the Matthiessen's rule [12]:

$$\frac{1}{\tau_{\text{tot}}} = \frac{1}{\tau_{AP}} + \frac{1}{\tau_{II}} + \frac{1}{\tau_{SR}} + \frac{1}{\tau_{DP}}. \tag{6}$$

The total mobility is determined by the transport lifetime: $\mu = e\tau/m^*$, with m^* as the in-plane effective mass of the carrier. The transport lifetime limited by some disorder is represented, via the scattering rate $1/\tau$, in terms of its autocorrelation function (ACF) by [12].

3. Autocorrelation functions for scattering mechanisms

With the use of the envelop wave function from Eq. (1), we are able to derive the autocorrelation functions for all possible low-temperature scattering mechanisms in strained 1S-doped square QWs in an analytic form (remote impurity-RI, surface roughness-SR, and misfit deformation potential-DP) [20].

At high temperatures, we are able to derive the autocorrelation functions for Phonon scattering. It has been shown [22] that phonons scattering plays an important role in limitting the electron mobility in III-V semiconducting compounds. The most important phonon scattering processes are (i) deformation potential acoustic, (ii) piezoelectric acoustic. In this paper we calculate only the temperature dependent part of the electron mobility. The momentum relaxation time due to acoustic-phonon scattering is given by:

$$\frac{1}{\tau_{AP}} = \sum_{k'} W(k,k')(1 - cos\vartheta), \tag{7}$$

where $W(k,k')$ is the transistion probability from state k to k' and ϑ is the scattering angle. In the temperature range in which we analyze the electron mobility, acoustic-phonon scattering can be consider as elastic. We approximate the phonon-distribution function N_q by $K_bT/\hbar\omega_q$. Then $W(k,k')$ is given by [23]

$$W(k,k') = \frac{2k_BT}{\hbar^2}\int_{-\infty}^{+\infty} dz \frac{|M(q)|^2}{\omega_q S^2(|k-k'|)} |I(q_z)|^2 \delta(E_k - E_{k'}), \tag{8}$$

where q_z is the wave vector normal to the heterointerface and $|M(q)|$ is the 3D scattering matrix element. $I(q_z)$ is the overlap integral by

$$I(q_z) = \int_{-\infty}^{+\infty} dz\zeta(z)exp(iq_z z)\zeta(z). \tag{9}$$

The scattering matrix elements for the deformation-potential and piezoelectric coupling are, respectively, given by

$$|M(q)|^2 = \frac{D^2\hbar\omega_q}{2c_l L^3}, \tag{10}$$

and

$$|M(q)|^2 = \frac{2\pi e^2 P^2 \hbar\omega_q}{\epsilon_0 q^2 L^3}. \tag{11}$$

Here ϵ_0 is the static dielectric constant $\epsilon_0 = 12.91$, e the elemental charge. The static screening factor at finite temperature has been calculated to be:

$$S(q) = 1 + \frac{2\pi e^2 F(q)\Pi(q)}{\epsilon_0 q}. \tag{12}$$

The screening form factor $F_S(q)$ takes account of the extension of the particle state along the growth direction, defined by

$$F_S(q) = \int_{-\infty}^{+\infty} dz \int_{-\infty}^{+\infty} dz' \zeta^2(z)\zeta^2(z')e^{-q|z-z'|}. \tag{13}$$

Here, we defined the following mathematical functions:

$$\gamma_n(x) = \left[\frac{1}{x} + \frac{(-1)^n x}{x^2 + n^2\pi^2}\right] \sinh x; \quad \omega_n(x) = \frac{(-1)^n n\pi}{x^2 + n^2\pi^2} \sinh x, \tag{14}$$

$F_S(t)$ defined by [21] with the wave function from Eq. (1), the calculation of Eq. (13) is lengthy, but straightforward with the aid of the functions γ_n and ω_n in Eqs. (14).
The static polarizability at finite temperature is given by [24]

$$\Pi(q, T, E_F) = \frac{1}{4k_B T} \int_0^{+\infty} \frac{\Pi(q, 0, \sigma)}{\cosh^2(\frac{E_F - \sigma}{2k_B T})} d\sigma, \tag{15}$$

where $\Pi(q, 0, \sigma)$, is the polarizability funtion at $T = 0K$, namely,

$$\Pi(q, 0, \sigma) = \frac{m*}{\pi\hbar^2}\{1 - \theta(q - 2k_F)(1 - (\frac{2k_F}{q})^2)^{1/2}\}. \tag{16}$$

Here $k_F = \sqrt{2m * E_F}/\hbar$ is the Fermi wave number. In the quantum limit, the temperature can be considered low enough so that $q < 2k_F$ becomes a very good approximation to acoustic phonon and it is very useful in phonon mobility calculations since it makes the factor in brackets of Eq. (17) to be unity. Also, due to the fact that significant effects of interface-phonons were never experimentally found, the electron-phonons non interactions occurring in the 2DEG are considerd identical to that of bulk GaAs.

4. Results and conclusion

4.1. Input material parameters

In this section, we apply the above-developed theory in order to understand the properties of low-temperature transport in remote doped square QWs. In Particular, we clarify the The main scattering mechanisms in Single-side modulation doped square quantum wells.

To illustrate all above-quoted scattering mechanisms, in the Figure 1, we consider a two-dimensional hole gas (2DHG) in the strained Ge channel of an $Si_{0.3}Ge_{0.7}$/Ge/$Si_{0.30}Ge_{0.7}$ QW. The two barriers are B-doped equally with a doping profile of $L_d = 100$ Å and $L_s = 100$ Å. As indicated, [9] for strained Ge on cubic [001] $Si_{0.3}Ge_{0.7}$, the potential barrier height for heavy holes is rather large: $V_0 \approx 270$ meV. Therefore, the assumption of infinite square QWs is plausibly accepted as in the previous theories. [13] For strained Ge, the out of-plane and in-plane

IOP Conf. Series: Journal of Physics: Conf. Series **865** (2017) 012004 doi:10.1088/1742-6596/865/1/012004

effective masses of the heavy hole are $m_z = 0.19\,m_e$, $m^* \approx 0.1\,m_e$. [17] The lattice mismatch of the QW is $\epsilon_{\|} = -0.01$. The elastic stiffness constants of the Ge channel are $c_{11} = 12.85$, $c_{12} = 4.83$, $c_{44} = 6.80$ in units of 10^{10} Pa, and the shear deformation potential constants are $b_s = -2.55$, $d_s = -5.50$ in eV [9, 25].

In the Figure 2, we we consider a two-dimensional electron gas (2DHG) in the strained GaAs channel of an $GaAs/AlGa_{1-x}As_x$ square QWs whose following physical parameters were taken elsewhere [26].

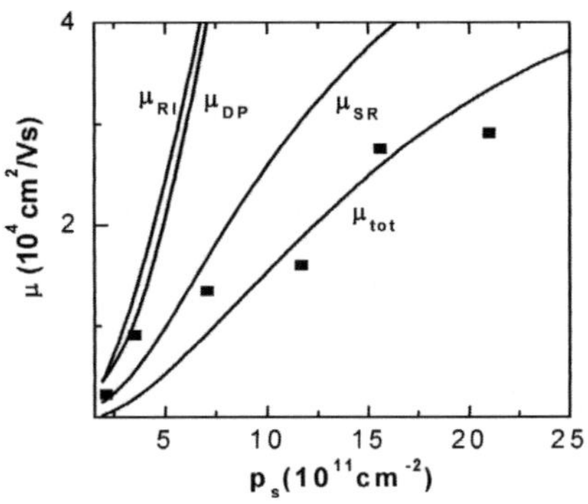

Figure 1. Partial and total mobilities of holes in a $Si_{0.33}Ge_{0.67}/Ge/Si_{0.33}Ge_{0.67}$ square QW with a chanel width $L = 75$ Å vs hole density p_s. The single-side modulation doping is with $L_d = 100$ Åand $L_s = 100$ Å. The 8K experimental data [6] are marked by squares.

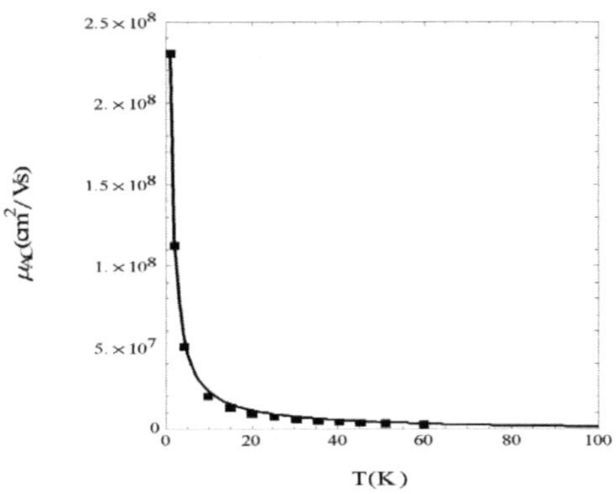

Figure 2. Acoustic-phonon partial mobility in a $GaAs/AlGa_{1-x}As_x$ square QW with a chanel width $L = 152$ Å vs temperature T. There common parameters are $L_d = 100$ Å , $L_s = 100$ Å . The experimental data [10] are marked by squares.

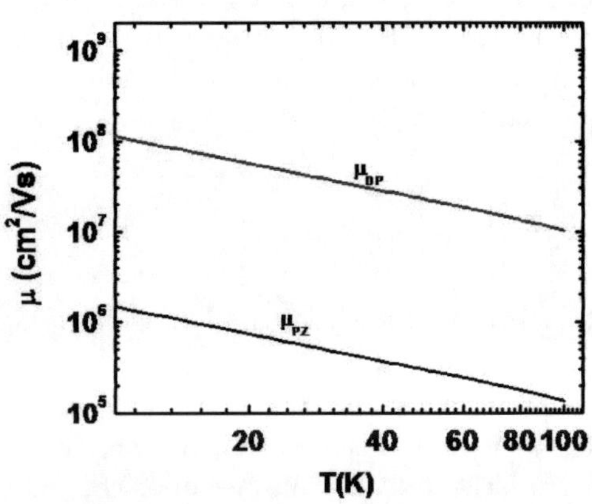

Figure 3. Acoustic-phonon (DP plus PZ coupling) partial mobility dependence on temperature for single-side doped square quantum well.

At high temperatures the carriers in the single-side doped square QWs are expected to experience the following scattering sources: acoustic phonons (Deformation potential-DP) and Piezoelectric coupling-PZ). From Figure 3, the key scattering mechanisms is acoustic phonons due to the piezoelectric coupling (PZ), acoustic phonons due to Deformation potential (DP) is in general less relevant.

4.2. Conclusions

From the lines thus obtained, we may draw the following conclusions:
(i) As clearly observed from Figure 1, the calculated overall mobility may reproduce very well the observed hole density p_s dependence of the 2DHG mobility. The key scattering mechanisms are surface roughness and misfit deformation potential, remote impurities are in general less relevant.
(ii) We calculated the dependences on temperature of acoustic-phonon scattering and the total mobility from Figure 2 and Figure 3. We show that phonon scattering plays an important role in limiting the electron mobility in square QWs. We analyzed the behavior of the mobility for temperature values up to 100K.
(iii) Our theory is able to well reproduce the recent experimental data about transport of electrons and holes in one-side doped square QWs.

References

[1] Schaffler F 1997 *Semicond. Sci. Technol.* **12** 1515
[2] Whall T E and Parker E H C 1998 *J. of Phys. D* **31** 1397
[3] Xie Y H, Monroe D, Fitzgerald E A, Silverman P J, Thiel F A and Watson G P 1993 *Appl. Phys. Lett.* **63** 2263
[4] Irisawa T, Miura H, Ueno T, Shiraki Y 2001 *Jpn. J. Appl. Phys.* **40** 2964
[5] Okuyama Y and Tokuda N 1989 *Phys. Rev. B* **40** 9744
[6] Irisawa T, Myronov M, Parker E H C, Nakagawa K, Murata M, Koh S, and Shiraki Y 2003 *Appl. Phys. Lett.* **82**, 1425
[7] Irisawa T, Miura H, Ueno T, Shiraki Y 2001 *Jpn. J. Appl. Phys.* **40** 2964
[8] Myronov M, Sawano K, and Shiraki Y 2006 *Appl. Phys. Lett.* **88** 252115
[9] Kahan A, Chi M and Friedman L 1994 *J. Appl. Phys.* **75** 8012
[10] Lima F M S, Qu Fanyao, Nunes O A C and Fonseca A L A 2001 *phys. stat.sol (b)* **1** 43-61
[11] Fang F F and Howard W E 1966 *Phys. Rev Lett.* **16** 797
[12] Gold A 1987 *Phys. Rev. B* **35** 10798
[13] Laikhtman B and Kiehl R A 1993 *Phys. Rev. B* **47** 10515
[14] Gold A 1988 *Phys. Rev. B* **38** 10798
[15] Ando T, Fowler A B and Stern F 1982 *Rev. Mod. Phys.* **54** 437
[16] Feenstra R M and Lutz M A 1995 *J. Appl. Phys.* **78** 6091
[17] Fischetti M V and Laux S E 1996 *J. Appl. Phys.* **80** 2234
[18] Lima F M S, Qu Fanyao, Nunes O A C and Fonseca A L A 2001 *Phys. stat. sol. (b)* **1** 225
[19] Quang D N, Tung N H, Hien D T, and Huy H A 2007 *Phys. Rev. B* **75** 073305
[20] Quang D N, Tung N H, Hien D T and Hai T T 2008 *J. Appl. Phys.* **104** 113711
[21] Quang D N, Tung N H 2008 *Phys. Rev. B* **77** 125335
[22] Rode D L 1975 *Semiconductors and Semimetals, edited by R. K. Willardson and A. C. Beer (Academic, New York, 1975), Vol. 10, Chap 1.*
[23] Price P J 1982 *Surf. Sci.* **113** 199
[24] Okuyama Y and Tukuda N 1989 *Phys. Rev. B* **40** 9744
[25] Van de Walle C G 1989 *Phys. Rev. B* **39** 1871
[26] Umansky V, De-Picciotto R and Heiblum M 1981 *Appl. Phys. Lett.* **39** 912

IOP Conf. Series: Journal of Physics: Conf. Series **865** (2017) 012005 doi:10.1088/1742-6596/865/1/012005

On the concentration dependence of metallic nano-particles in enhanced Forster resonance energy transfer

N M Hoa[1], T H Nhung[2], T T Thao[3], L T Theu[4], N A Viet[2]

[1] Hue University of Medicine and Pharmacy, Hue, Viet Nam
[2] Institute of Physics, Vietnam Academy of Science and Technology, 10 Dao Tan, Ba Dinh, Ha Noi, Viet Nam
[3] Post and Telecommunications Institute of Technology, Ha Noi, Viet Nam
[4] Hanoi Pedagogical University 2, Vinh Phuc, Viet Nam

Email: minhhoa2806@gmail.com

Abstract. The enhanced Forster resonance energy transfer is obtained in recent experiments. One of the most important experimental facts is the observed strong enhancement of donor fluorescence with increasing concentration of metallic nanoparticle acceptors in the first stage. We propose three types of energy transfer mechanisms (Forster FRET, surface SET, and Coulomb CET) and predict simple relations for their dependencies on the concentration of metallic nano-particles. The competition of these mechanisms leads to the transfer intensity dependence on the concentration dependence of metallic nano-particles. Using these assumptions, we expect and argue about the compatibility between the experimental results.

1. Introduction

Plasmon effects of metal nanoparticles have many applications for a variety of application areas such as: chemical analysis and catalysis, detection of biomolecules, pharmaceutics, diagnosis, imaging and therapy, discovering the contamination by heavy metal ions, analyzing food additives and contamination [1-10].

The Foster Resonance Energy Transfer (FRET) is a process of energy transmission through dipole-dipole interaction, from a chemical molecule (donor - D) to the acceptor molecule (acceptor - A), but here the donor emission band must overlap the acceptor's absorption bands. Currently, FRET effects are widely applied in the fields of chemistry, biology such as stem cell research, as well as studies of proteins, and biometric applications.

In recent decades, metal nanostructures are famous by the attractive optical properties of the electromagnetic spectrum from UV to MIR spectral region. For example, they are used to enhance the optical signal in the fluorescence, Raman spectrum, IR spectrum, and increase the contrast in nuclear magnetic resonance (MRI). Its nature is fundamentally enabled by the surface plasmon resonance-based localized collective oscillations of electron cloud with metal nanoparticles.

IOP Conf. Series: Journal of Physics: Conf. Series **865** (2017) 012005 doi:10.1088/1742-6596/865/1/012005

FRET between two molecular donors (D - usually a fluorophore) and acceptor, in which used acceptor is a metal nanoparticle is an important physical method for investigation biological systems with applications in medical treatment and drug discovery today [12,13].

Within the limits of the distance between A and D equal to 10 nm, we have given a new reasonable model with mixing mechanism: FRET-mechanism for small size and SET-mechanism for large size metal nanoparticles. Even with a research by electronic excitation transfer (CET) from a segment of polyfluorene to tetraphenyl porphyrin, Bagchi et al. have show the R^{-2} dependence rule [16].

We propose a GRET model [22] with three types of energy transfer mechanisms (Forster FRET, surface SET, and Coulomb CET) and predict simple relations for their dependencies on the concentration of metallic nano-particles. The competition of these mechanisms leads to the transfer intensity dependence on the concentration dependence of metallic nano-particles.

In this paper we examine this General resonance energy transfer (GRET) model and plasmonic effects of nanoparticles for more recently obtained experimental data. We will show that our GRET model gives a quite good agreement with experimental data, and could explain the enhanced fluorescence intensity in the presence of gold particles with their concentration dependence.

2. Theoretical Model

The first paragraph after a heading is not indented (Bodytext style). We propose three types of energy transfer mechanisms (Forster FRET, surface SET, and Coulomb CET) and predict simple relations for their dependencies on the concentration of metallic nano-particles [22].

Denote the FRET, SET, and CET efficiencies as given by

$$E_{\mathrm{FRET}}=\frac{1}{1+\left(R/R_{\mathrm{FRET}}\right)^6}\ , \tag{1}$$

$$E_{\mathrm{SET}}=\frac{1}{1+\left(R/R_{\mathrm{SET}}\right)^4}\ , \tag{2}$$

$$E_{\mathrm{CET}}=\frac{1}{1+\left(R/R_{\mathrm{CET}}\right)^2}\ , \tag{3}$$

where R_{FRET}, R_{SET} and R_{CET} are the FRET, SET and CET radii, respectively. $R=\left|\vec{r}_A-\vec{r}_D\right|$ is the distance between the A and the D. At $R=R_{\mathrm{FRET}}$, $R=R_{\mathrm{SET}}$, and $R=R_{\mathrm{CET}}$, half of excitation energy of D is transferred to nano-metallic particle A.

Denote $D\left(\vec{r}_A,\vec{r}_D;\omega\right)$ the Green function. The function M is defined as

$$M(\omega)=\left|\overrightarrow{G_D}(\omega)\cdot D\left(\overrightarrow{r_A},\overrightarrow{r_D};\omega\right)\cdot\overrightarrow{G_A}(\omega)\right|^2\ . \tag{4}$$

We have [17,20]

$$M(\omega,\langle R\rangle)=\frac{1}{\left(4\pi\varepsilon_0\right)^2}K^2G_A(\omega)^2G_D(\omega)^2\left[\frac{3}{\langle R\rangle^6}+\frac{(\omega/c)^2}{\langle R\rangle^4}+\frac{(\omega/c)^4}{\langle R\rangle^2}\right] \tag{5}$$

where $K^2=[0,1]$ is the orientation factor of effective dipoles and $\langle R\rangle$ is the average D-A distance. In the equation (5) there are just 3 contributions given by FRET, SET, and CET-mechanisms respectively.

We assume a simple relation between the average donor-acceptor distance $\langle R\rangle$ and NPs concentration n as $\langle R\rangle \sim (n_A + n_D)^{-1/3} \approx n^{-1/3}$.

We have

$$M(\omega,\langle R\rangle)=\frac{1}{\left(4\pi\varepsilon_0\right)^2}K^2G_A(\omega)^2G_D(\omega)^2\left[An^2+Bn^{4/3}+Cn^{2/3}\right]\ , \tag{6}$$

where A,B, and C are some constants and n_A, n_D are concentration of A and D, respectively.

The competition of these mechanisms: give more energy to the donor via CET mechanism ($c > 0$) and take the energy from a donor via FRET and SET mechanisms ($a < 0,\ b < 0$). With the usual linear optic condition $(G_AG_D)^2 \sim n$, the normalized intensity I is

$$I(n)=I_0+an^3+bn^{7/3}+cn^{5/3}\ , \tag{7}$$

IOP Conf. Series: Journal of Physics: Conf. Series **865** (2017) 012005 doi:10.1088/1742-6596/865/1/012005

where *a,b*, and *c* are some constants, I_0 is the normalized intensity at $n \approx 0$.

The form of normalized luminescence intensity I (C-F-S) is plotted in Figure 1 with the assumption that the contribution of CET is positive (c > 0), the contributions of FRET and SET are negative ($a < 0$, $b < 0$). At small frequency ω, the intensity always decreases with increasing the density of NPs concentration n. At high frequency ω, the intensity first increases with increasing n until some certain value n_{max}, then decreases after this point. This behavior of the luminescence intensity I versus the density of NPs concentration n was observed in the experiment.

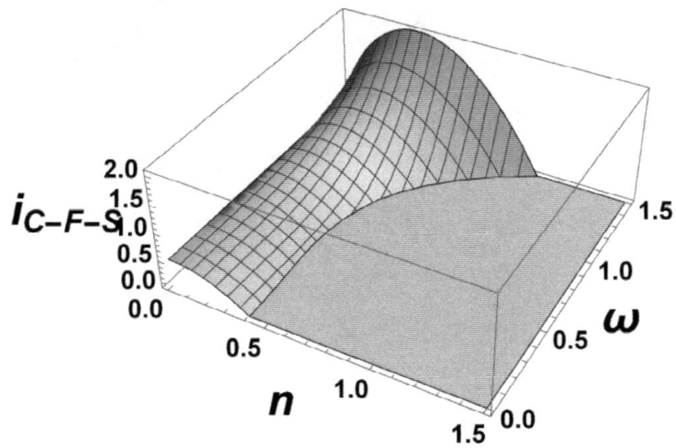

Figure 1. The form of normalized luminescence intensity I (C-F-S) with the assumption that the contribution of CET is positive ($c > 0$), the contributions of FRET and SET are negative ($a < 0$, $b < 0$). This behavior of the luminescence intensity I versus the density of NPs concentration n is the same as in experimental observations.

The average distance r-dependence and frequency ω-dependence of the ratio of CET contribution versus FRET contribution are presented in Figure 2a, and the ratio of SET contribution versus FRET contribution is presented in Figure 2b.

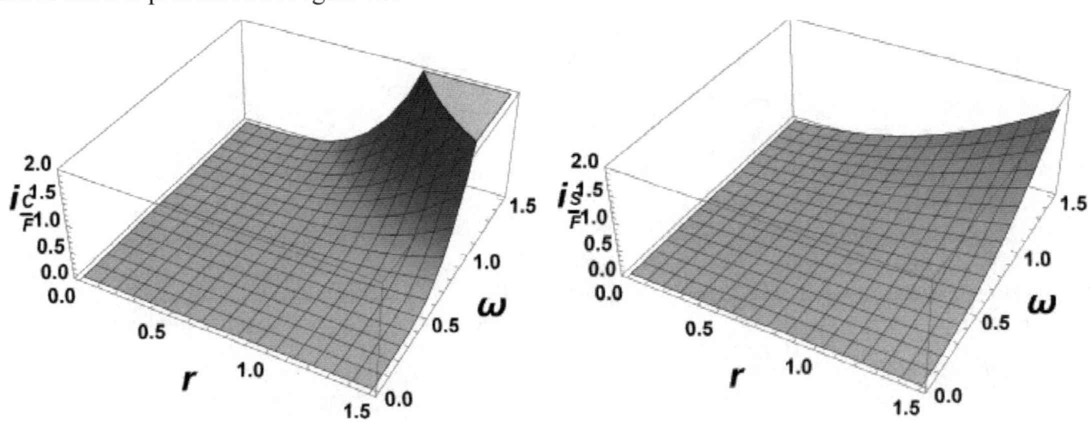

(a) (b)

Figure 2. The ratio of CET contribution versus FRET contribution (a), and the ratio of SET contribution versus FRET contribution (b).

This is entirely consistent with our theoretical models given above.

3. Comparison with experiment

We investigated the resonance energy transfer between fluorophores (D) and the metal nanoparticles (A). As a result, it caused the extinction or fluorescence enhancement. The experiments were conducted at room temperature and at 532 *nm* excitation for a mixture of D-A pairs in aqueous solution. Using the JASCO-V570-UV-Vis-NIR spectrometer were measured absorption spectra, a Cary Eclipse spectrofluorometer were recorded. The fluorescence spectra, transmission and scanning electron microscopes (TEM, JEM 1011 and SEM, Hitachi S-480) were used to determine the shape, size of particles. The energy transfer experiments were carried out for two types of D-A pairs: CdTe QDs – GNPs, RBDSNPs – GNPs (20 nm and 100 nm GNPs).

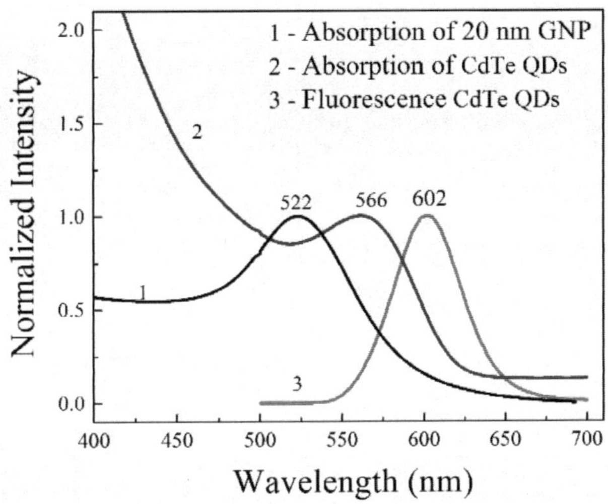

Figure 3. Absorption and fluorescence spectra for CdTeQDs-GNPs.

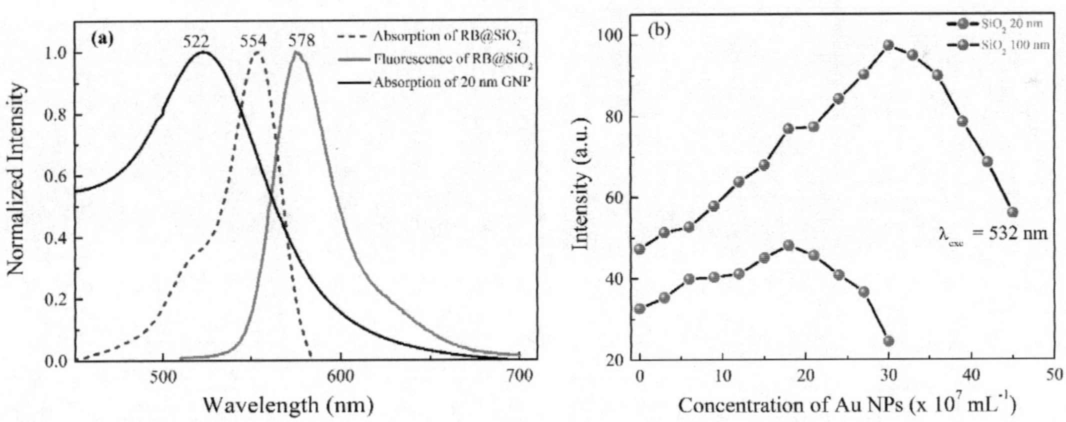

Figure 4. Absorption, fluorescence spectra (a) and donor fluorescence intensity vs acceptor concentration of D-A pair (b) between RBDSNPs – (20 and 100) nm GNPs.

Using the model developed in the previous part, we calculate the fluorescence dependence on the concentration of gold nanoparticles. The theoretical fitting results have a quite good agreement with the above experiments. This model has also been tested in previous report [22].

From Figures 3 to 6 we can see that at a small frequency $\omega<\omega_0$, where ω_0 is some certain frequency, the luminescence intensity always decreases with increasing the density of NPs concentration *n*. At higher frequency $\omega<\omega_0$, the intensity first increases with increasing *n* until some certain value $n_{\max}$, then decreases after this point. That behavior was mentioned in the above our model theory and also was observed in the experiment.

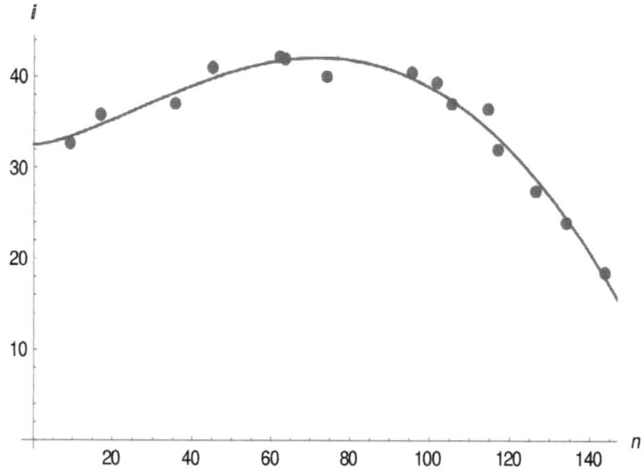

Figure 5. Results of theoretical models. The blue curve is a fit with parameters: $a = -8.2785\text{x}10^{-7}$, b = $-1.088\text{x}10^{-3}$, $c = 2.6794\text{x}10^{-2}$, $I_0 = 32.48$, and the experiment data are shown in red cycles.

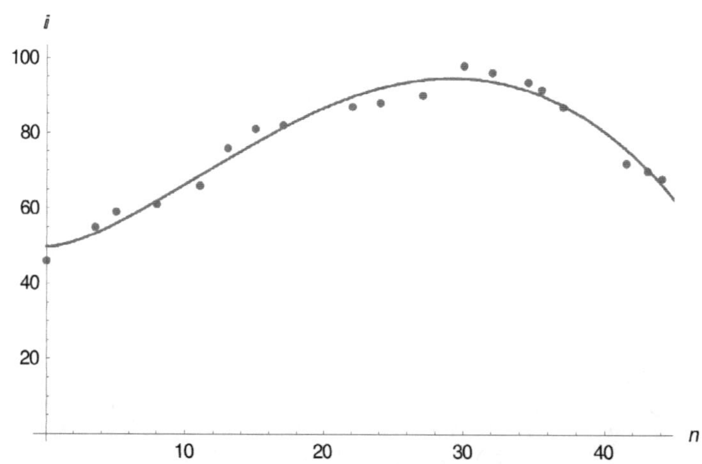

Figure 6. Results of theoretical models with GNPs (Au 20 nm). The blue curve is a fit with parameters $a = -3.9947\text{x}10^{-4}$, $b = -3.5438\text{x}10^{-3}$, $c = 0.5343$, $I_0 = 49.6958$, and the experiment data are shown in red cycles.

So we can confirm the mode of general resonance energy transfer is a mix of 3 mechanisms FRET, SET and CET. The dependence of each mechanism depends on the distance between D and A that we have surveyed.

4. Conclusion

Resonance energy transfer between fluorescent and fluorescent metal surface enhanced fluorescence has many applications in various fields. The results of this paper propose a theoretical model for calculating the fluorescence intensity versus concentration of gold nanoparticles.

We propose a GRET model with three types of energy transfer mechanisms (Forster FRET, surface SET, and Coulomb CET) and predict simple relations for their dependencies on the concentration of metallic nano-particles. The competition of these mechanisms can be explained as: giving more energy to donor via CET mechanism (c > 0) and taking the energy from donor via FRET and SET mechanisms (a < 0, b < 0) lead to the transfer intensity dependence on the concentration dependence of metallic nano-particles.

This results is quite important contribution in the study of fluorescent enhancement. Our model succeeds in explaining how the fluorescence intensity depends on the concentration of gold particles in experiments with CdTeQDs-GNPs, 20 and 100 nm RBDSNPs-GNPs. In the future, we can develop antenna models describing the transceiver enhanced fluorescence signal between D and A.

Acknowledgements
This research was financially supported by the Vietnam National Foundation for Science and Technology Development (NAFOSTED) under grant number 103.01-2015.42.

References

[1] Xie W and Schlücker S 2004 *Reports on Progress in Physics* **77** 116502
[2] Chen X-J, Cabello G, Wu D-Y and Tian Z-Q 2014 *Journal of Photochemistry and Photobiology C: Photochemistry Reviews* **21** 54-80
[3] Schlücker S 2014 *Angew. Chem. Int. Ed.* **53** 4756-4795
[4] Cialla D, Pollok S, Steinbrücker C, Weber K and Popp J 2014 *Nanophotonics* **3** 383–411
[5] Ma W, Kuang H, Xu L, Ding L, Xu C, Wang L and Kotov N A 2013 *Nat. Commun.* **4** 2689
[6] McAughtrie S, Faulds K and Graham D 2014 *Journal of Photochemistry and Photobiology C: Photochemistry Reviews* **21** 40-53
[7] Dinh T V, Liu Y, Fales A M, Ngo H, Wang H-N, Register J K, Yuan H, Norton S J and Griffin G D 2015 *Wiley Interdisciplinary Reviews: Nanomedicine and Nanobiotechnology* **7** 17-33
[8] Li D-W, Zhai W-L, Li Y-T and Long Y-T 2014 *Microchim. Acta* **181** 23-43
[9] Alvarez-Puebla R A and Liz-Marzán L M 2012 *Angew. Chem. Int. Ed.* **51** 11214-11223
[10] Zheng J and He L 2014 *Comprehensive Reviews in Food Science and Food Safety* **13** 317-328
[11] Keller E L, Brandt N C, Cassabaum A A and Frontiera R R 2015 *Analyst* **140** 4922-4931
[12] Hussain S A. et al 2009 *Introduction to FRET,* arXiv preprint arXiv:0908.1815
[13] Chen G, Song F, Xiong X, Peng X 2013 *Ind. Eng. Chem. Res* ***52*** 11228–11245
[14] Wong K F, Bagchi B, Rossky P J 2004 *J. Phys. Chem. A* **108** 5752
[15] Vincent R, Carminati R 2011 *Phys. Rev. B* **83** 165
[16] Singh H, Bagchi B 2005 *Curr. Sci.* **89** 1710
[17] Breshike C J, Riskowski R A, and Strouse G F 2013 *J. Phys. Chem. C* **117** 23942
[18] Ha C V, Nga D T, Viet N A, Nhung T H 2015 *J. Opt. Comm.* **353** 4955
[19] Wong, K F, Bagchi B, and Rossky, P J 2004 *J. Phys. Chem. A* **108** 5752–5763
[20] Tuncay O, Pedro L H, Evren M, Onur A Sedat N, Ilkem O, Qing Z, Qihua X, and Hilmi V D 2013 *Nano Lett.* **13** 3065−3072
[21] Andrews D L, Bradshaw D S 2004 *Eur. J. Phys.* **25** 845858
[22] Hoa N M, Ha C V, Nga D T, Lan N T, Nhung T H, and Viet N A 2016 *J. Phys.: Conf. Ser.* **726** 012009

IOP Conf. Series: Journal of Physics: Conf. Series **865** (2017) 012006 doi:10.1088/1742-6596/865/1/012006

Some applications using the connection between q-deformed harmonic oscillator and symmetric and asymmetric potentials

Ngo Gia Vinh[1], Man Van Ngu[2], Nguyen Tri Lan[3], Luu Thi Kim Thanh[4], Nguyen Thi Dung[5] and Nguyen Ai Viet[3]

[1]Bac Ninh Department of Education and Training, Bac Ninh, Viet Nam
[2]Hung Yen Industrial College, Tu Son, Bac Ninh, Viet Nam
[3]Institute of Physics, Vietnam Academy Science and Technology, 10 Dao Tan, Ba Dinh, Hanoi, Viet Nam
[4]Hanoi Pedagogical University No.2, Nguyen Van Linh Str., Vinh Phuc, Viet Nam
[5]Bac Ninh Teacher Training College, 12A Binh Than, Dai Phuc Ward 10, Bac Ninh, Viet Nam

E-mail: ngogiavinh@bacninh.edu.vn

Abstract. In our previous article, the connections between q-deformed harmonic oscillator and the two types of asymmetric (Morse-like) and symmetric (inverse square cosine form) potentials have been investigated. The use of these relations in an inverse way to investigate the properties of q-deformed harmonic oscillators has been proposed. In this work, we explore the possibility of using this approach to study some real physical systems, such as diatomic molecules, phonon, etc.

1. Introduction

Deformed Heisenberg algebras with q-deformed harmonic oscillator have been a subject of intensive investigation and have many useful applications in physics and chemistry [1–7]. The method of q-deformed quantum mechanics was based on the Heisenberg commutation relations for bosons. The main parameter of this method is a real number $q \in [0, 1]$, called deformation parameter, and $q = 1$ is corresponding to the case of in-deformed normal harmonic oscillators.

In the atomic and molecular physics, the interaction between atoms in diatomic and even in polyatomic molecules is usually described by the Morse potential [8–12]. In algebraic approach, the Morse potential can be written in terms of the generators of $SU(2)$ group. The quantum relation between q-deformed harmonic oscillator and the Morse potential was considered in [10], where the anharmonic vibrations in the Morse potential have been described as the levels of q-deformed harmonic oscillator. The extended $SU(2)$ model (q-Morse potential) has been also developed to compare with phenomenological Dunham expansion and experimental data for numbers of diatomic molecules [10].

In our previous works [14–17], considering deformed algebra as mathematical object and atomic effective potential as physical model, we proposed the new representations for q-deformed harmonic oscillator on the base of the physics model potential. In this representation, the potential for normal harmonic oscillation is parabolic with infinity equal-step energy levels. The potentials for q-deformed harmonic oscillation are anharmonic Morse potential (anharmonic

IOP Conf. Series: Journal of Physics: Conf. Series **865** (2017) 012006 doi:10.1088/1742-6596/865/1/012006

representation) or symmetric Pöschl–Teller potential (symmetric representation). In these case, energy spectra have finite in-equal steps, which are characterized by a maximal step number n_{max}.

In this work, using our developed approach we study some real physical problems: asymmetric representation for diatomic molecules and symmetric representation for phonon in nano structures.

2. Asymmetric representation of q-deformed harmonic oscillator

The Morse potential has the form

$$V(x) = D\left(1 - e^{-k(x-x_0)}\right), \tag{1}$$

where D is the depth, k is the strength, ω is the characterized frequency, and x_0 is the equilibrium position of the Morse potential.

The energy spectrum for Morse potential is

$$E(n) = \hbar\omega\left[\left(n+\frac{1}{2}\right) - \frac{\varepsilon}{2}\left(n+\frac{1}{2}\right)^2 + C_M\right], \tag{2}$$

where n is the vibrational quantum number.

The energy spectrum becomes quadratic if the higher order contribution C_M is neglected. In equation (2), the energy levels are represented by a system of parallel lines that are not equidistant.

$$\frac{E(n)}{\hbar\omega} = \left(n+\frac{1}{2}\right) - \frac{\varepsilon}{2}\left(n+\frac{1}{2}\right)^2. \tag{3}$$

The energy levels of the Morse potential and the corresponding harmonic potential are plotted in Figure 1.

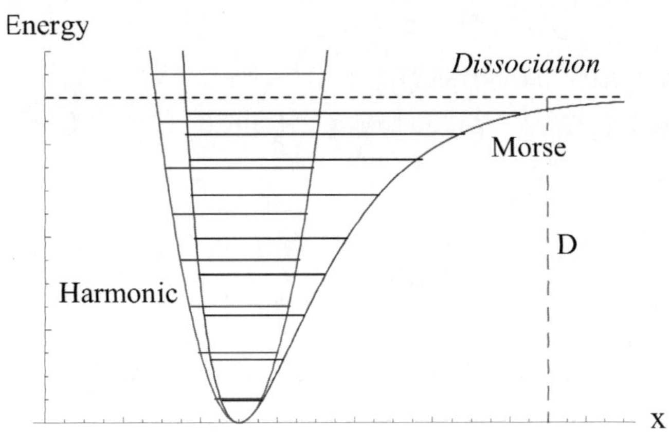

Figure 1. The energy levels Morse potential and harmonic potential.

IOP Conf. Series: Journal of Physics: Conf. Series **865** (2017) 012006 doi:10.1088/1742-6596/865/1/012006

3. Asymmetric representation and diatomic molecules

The energy spectrum of diatomic molecules with phenomenological Dunham quadratic expansion is

$$E(n) = hc\omega_e\left(n + \frac{1}{2}\right) - hc\omega_e x_e\left(n + \frac{1}{2}\right)^2 + ..., \tag{4}$$

where $\omega_e = \frac{\omega}{2\pi c}$, ω is frequency, c is the speed of light in vacuum and n is the vibrational quantum number. The Dunham vibrational molecular ω_e, $\omega_e x_e$ are obtained by fitting the potential curve to the experimental spectral data, $\omega_e x_e \ll \omega_e$ [10].

If the expansion (4) is truncated to the quadratic term, one obtains essentially the discrete spectrum of the Morse potential

$$E(n) = hc\omega_e\left[\left(n + \frac{1}{2}\right) - x_e\left(n + \frac{1}{2}\right)^2\right], \tag{5}$$

or

$$\frac{E(n)}{hc\omega_e} = \left(n + \frac{1}{2}\right) - x_e\left(n + \frac{1}{2}\right)^2. \tag{6}$$

Comparing the coefficients in the expressions (3) and (6), we can express the parameter-set of the Morse potential via the Dunham parameters, the depth D is

$$D = \frac{\omega_e}{4x_e}, \tag{7}$$

and the strength k equals

$$k = \sqrt{\frac{4\pi c}{\hbar}\mu\omega_e x_e}. \tag{8}$$

Setting $k_0 = \sqrt{\frac{4\pi c}{\hbar}\mu\omega_e}$ we get

$$k = k_0\sqrt{x_e}. \tag{9}$$

The q-deformed parameter can be expressed via x_e

$$q = 1 - 2x_e, \tag{10}$$

and the values of largest number n_{max} is

$$n_{max} = \left[\frac{1}{1-q}\right]. \tag{11}$$

The experimental value of the Dunham constant x_e of diatomic molecules are taken from [10], and the corresponding our calculated values of q, D/ω_e, k/k_0, n_{max} are presented in Table 1 and in Figures 2-5.

Diatomic molecules	$x_e.10^{-2}$	q	$\frac{D}{\omega_e}$	$\frac{k}{k_0}$	n_{max}
AgBr	2.4613	0.950774	10.1572	0.1569	20
HF	2.1764	0.956472	11.4869	0.1475	22
AgCl	2.1352	0.957296	11.7085	0.1461	23
NaH	1.74179	0.965164	14.3531	0.1320	28
AlH	1.73217	0.965357	14.4328	0.1316	28
BO	0.884985	0.982300	28.2491	0.0941	68
CO	0.617701	0.987646	40.4727	0.0786	80
SnO	0.528665	0.989427	47.2889	0.0727	94
KCl	0.32143	0.993571	77.7774	0.0567	155
KBr	0.30303	0.993939	82.5001	0.0550	165

Table 1. The values of the parameters of diatomic molecules.

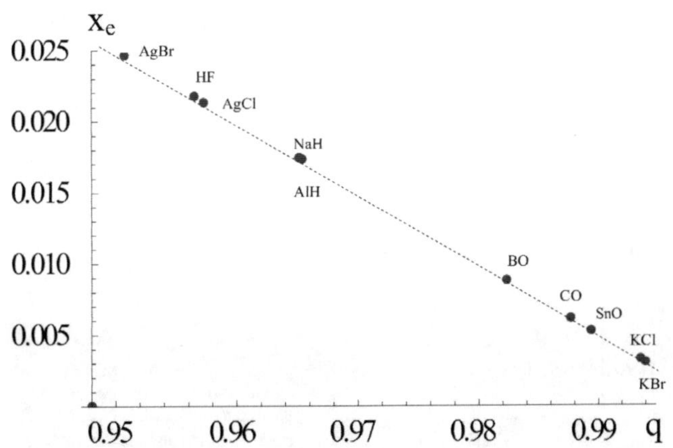

Figure 2. The dependence of x_e on q of some diatomic molecules.

From Figures 2-5 we can conclude that among the investigated here diatomic molecules, the most deformed diatomic molecule is AgBr with the deformation parameter q_{AgBr} =0.950774, and the less deformed diatomic molecule is KBr with the deformation parameter q_{KBr} =0.993939. All real diatomic molecules have very small deformation with $0.95 < q < 1$.

4. Symmetric representation of q-deformed harmonic oscillator

In the work [17] we have studied the deformation parameter q through Pöschl – Teller potential

$$V_s(x) = -\frac{U_0}{\cosh^2(\alpha x)}. \tag{12}$$

This potential is a symmetric anharmonic potential and has in-equal-step quadratic energy spectrum with the largest number n_{max} as above.

The energy levels of symmetric potential and harmonic potential are shown in Figure 6.

In the next session, we use this potential to study the phonon problem.

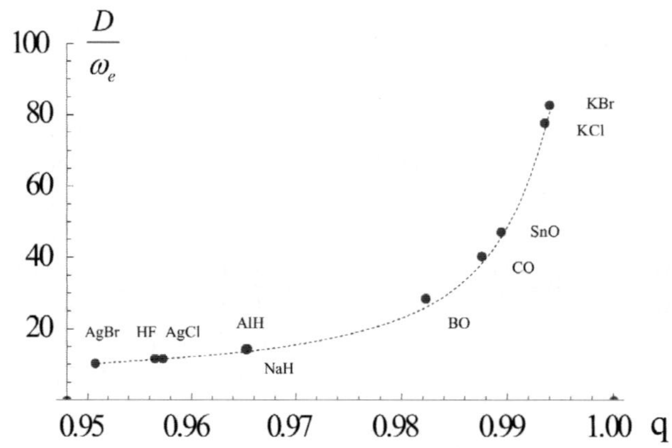

Figure 3. The dependence of $\frac{D}{\omega_e}$ on q of some diatomic molecules.

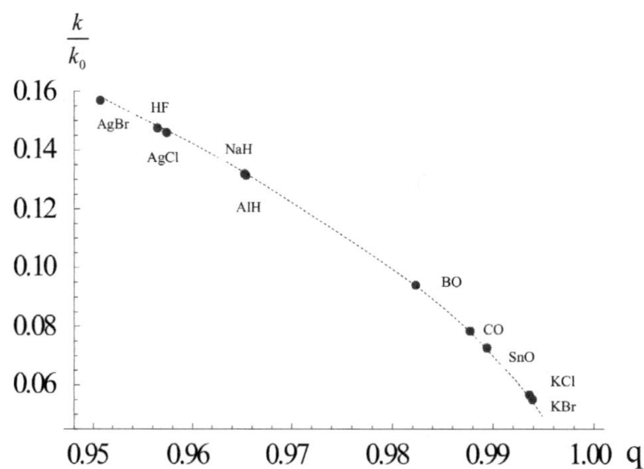

Figure 4. The dependence of $\frac{k}{k_0}$ on q of some diatomic molecules.

5. Symmetric representation and phonon in 1D atomic chain

It is well known that, phonon in the solids can be treated as a boson gas. For simplicity we take the case of 1D atomic lattice and study the deformation effect of phonon in this system.

For 1D atomic chain, $n_{max} = L/a$ where a is the inter-atomic distance, L is the chain-length. Usually in real crystals $a \simeq 5$Å, and if L is very short (of few nanometers) we have only small n_{max}.

We apply symmetric representation of q–deformed harmonic oscillator for phonon in the 1D atom chain with standard relation between the deformation parameter q and n_{max}: $n_{max} = \left[\frac{1}{1-q}\right]$.

For the case of long chains (a large-sized structures) $L \gg a$, $n_{max} \to \infty$, $q \to 1$, deformation effect is very small and we can use the Einstein-Debye cut-off frequency: $\omega_E = \omega_0$, $n_{max} = \frac{L}{a} \gg 1$.

For the inverse case of very short chain $n_{max} = L/a = [1/(1-q)]$. For example $L = 5nm$,

IOP Conf. Series: Journal of Physics: Conf. Series **865** (2017) 012006 doi:10.1088/1742-6596/865/1/012006

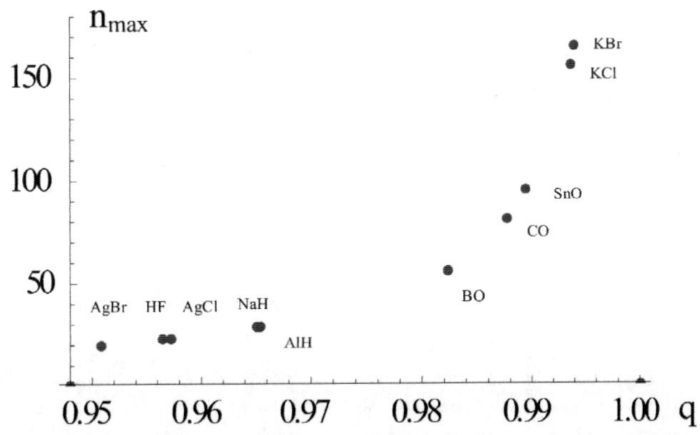

Figure 5. The dependence of n_{max} on q of some diatomic molecules.

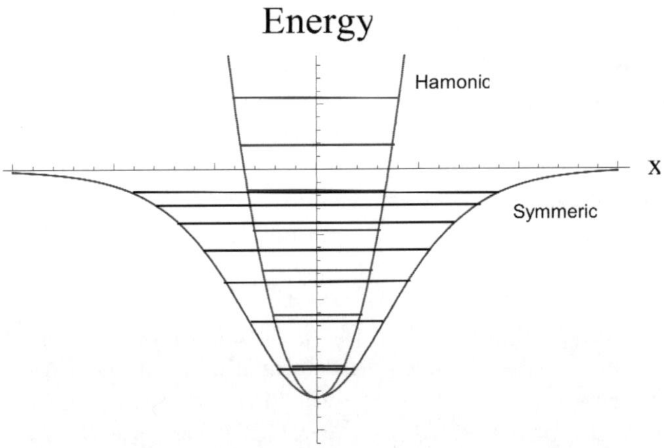

Figure 6. Energy levels of symmetric and harmonic potential.

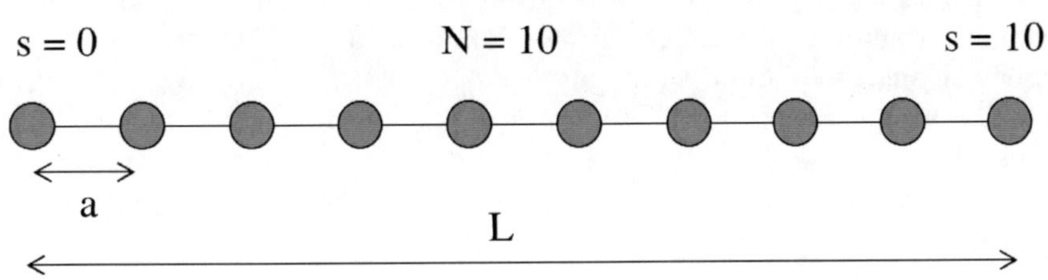

Figure 7. Model of 1D atomic lattice.

$n_{max} = 10$, the deformation parameter is $q \sim 0.9$. We conclude that for nano size chains the deformation effect is important.

6. Conclusion

In our previous works considering deformed algebra as mathematical object and atomic effective potential as physical model, we proposed a new representation for q-deformed harmonic oscillator on the base of a physical model potential. In this representation, the potential for normal

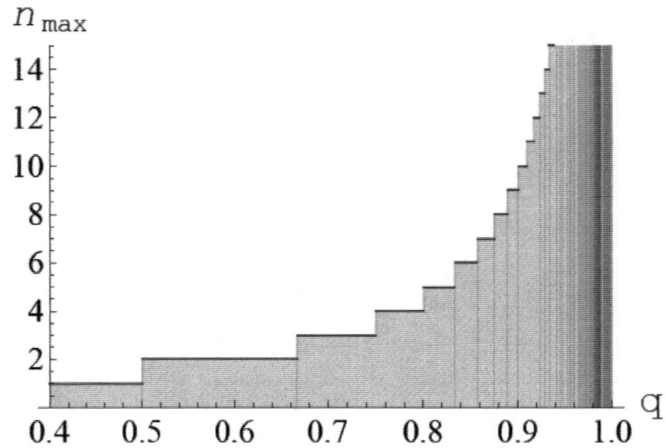

Figure 8. The values of the largest numbers n_{max} depending on deformation parameter q.

harmonic oscillation is parabolic with infinity equal-step energy levels. The potentials for q-deformed harmonic oscillation are asymmetric Morse potential (asymmetric representation) or symmetric Pöschl–Teller potential (symmetric representation). In these cases, energy spectra have finite in-equal steps, which are characterized by a maximal step number n_{max}.

In this work, using our developed approach we study some real physical problems: asymmetric representation for diatomic molecules and symmetric representation for phonon in nano structures.

We prove that the anharmonic potential deformation representation works well for the case of diatomic molecules. The physical parameters for diatomic molecules depending on deformation parameter q are plotted in Figures 2-5 with well determined curves. From those figures we can see that among the investigated here diatomic molecules, the most deformed diatomic molecule is AgBr with the deformation parameter q_{AgBr} =0.950, and the less deformed diatomic molecule is KBr with the deformation parameter q_{KBr} =0.993. All investigated real diatomic molecules have very small deformation of $0.95 < q < 1$.

We applied the symmetric potential deformation representation for phonon problem in very short atomic chain of nanometer length. We found that for 1D atomic nano chains the deformation effect of phonon is important.

Our developed approach with deformed potentials should be useful in investigating other physics problems. This will be the subject of our next work.

Acknowledgments
This research is funded by Vietnam National Foundation for Science and Technology Development (NAFOSTED) under grant number 103.01-2015.42.

References
[1] Eremin V V and Meldianov A A 2008 *arXiv:0810.1967v1*
[2] Mirzrahi S S, Camargo J P and Dodonov V V 2004 *J. Phys.* A **37** 3707
[3] Abe S 2009 *Phys. Rev.* E **79** 041116
[4] Lavagno A and Swamy P P 2010 *Physica* A **389** 993
[5] Algin A, Senay M 2012 *Phys. Rev.* E **85** 041123
[6] Andrade F M and Silva E O 2013 *Phys. Lett.* B **719** 467
[7] Tkachuck V M 2013 *Phys. Rev.* A **86**, 062112
[8] Angelova M and Franck A 2005 *Phys. Atom. Nuclei* **68** 1689-1697
[9] Dong S H, Lemus R and Frank A 2002 *J. Quant. Phys.* **86** 433

[10] Angelova M, Dobrev V and Frank A 2004 *Eur. Phys. J.* D **31** 27
[11] Kaplan I G 2003 *Handbook of Molecular Physics and Quantum Chemistry* (Wiley)
[12] Lima E F and Hornos J E M 2005 *J. Phys.* B **38** 815
[13] Landau L D and Lifshitz E M 1991 *Course of Theoretical Physics* Vol. 3
[14] Vinh N G, Ngu M V, Lan N T, Viet N A and Thanh L T K 2015 *J. Phys.: Conf. Ser.* **627** 012022
[15] Ngu M V, Vinh N G, Lan N T, Viet N A and Thanh L T K 2015 *J. Phys.: Conf. Ser.* **627** 012021
[16] Vinh N G, Ngu M V, Lan N T, Viet N A and Thanh L T K 2016 *J. Phys.: Conf. Ser.* **726** 012018
[17] Ngu M V, Vinh N G, Lan N T, Viet N A and Thanh L T K 2016 *J. Phys.: Conf. Ser.* **726** 012017

IOP Conf. Series: Journal of Physics: Conf. Series **865** (2017) 012007 doi:10.1088/1742-6596/865/1/012007

On the theory of three types of polaritons (phonon, exciton and plasmon polaritons)

Duong Thi Ha[1,2] , **Dinh Thi Thuy**[3] , **Vo Thi Hoa**[4] , **Tran Thi Thanh Van**[2] , **Nguyen Ai Viet**[5]

[1] Thai Nguyen University of Education, Thai Nguyen, Vietnam
[2] Graduate University of Science and Technology, Vietnam Academy of Science and Technology, 18 Hoang Quoc Viet, Cau Giay, Ha Noi, Vietnam
[3]Thai Binh University of Medicine, Thai binh, Vietnam
[4] Quang Nam university, Quang Nam, Vietnam
[5]Institute of Physics, Vietnam Academy of Science and Technology, Ha Noi, Vietnam

E-mail: duongha.sp@gmail.com

Abstract. We have investigated the similarities and difference between three well-known types of polaritons: phonon polariton, exciton polariton and surface plasmon polariton. For first two types (phonon polariton and exciton polariton) the interaction between photon and media can be expressed via a longitudinal-transversal splitting (LT-splitting), while for third type of polariton (surface plasmon polariton) via the boundary condition. Considering an analogy of these three types of polaritons, an effective LT-splitting was introduced for surface plasmon polariton. We discuss a possible existence of an evanescent state in the band gap of polaritons. Finally, the Nambu broken symmetry theory and Anderson-Higgs mechanism are discussed for lower branch of these polaritons.

1. Introduction

In recent years, the study of polariton has a great attention on both experimental and theoretical sides. The concept of polaritonics is recently introduced to the studies of polariton and its applications. The frequency spectra of polaritonics ranges from hundreds of gigahertz to several terahertz, which is the frequency gap between electronic and photonics. Therefore, polaritonics plays an important role in enabling advanced signal processing and spectroscopy application [1], allows observation of polariton condensation specially at room temperature [2-4], stimulates scattering [5], quantized vortices and super fluid behavior [6-7]. Besides that, polaritons can be applied in ultrafast optical switches, ultralow threshold polariton lasers and light emitting diodes [8-10].

Polariton can be simply defined as the strong coupling of photon with another quasiparticle. One can distribute polariton in different types: phonon polariton, exciton polariton and plasmon polariton.

The phonon polariton concept was introduced by Kun Huang (1951) when he studied the coupling of photon with phonon. In 1958, Hopfield studied the coupling of photon with exciton and suggested the concept of exciton polariton. Since then, the theoretical models of exciton polariton and phonon polariton have mainly focused on bulk polariton and neglected surface polariton lines which are in LT gap. In the case of plasmon polariton, most of works have focused

on surface plasmon polariton with frequency less than surface plasmon frequency. The range of frequency between bulk plasmon and surface plasmon is considered as the forbidden gap.
Recently, the existence of novel evanescent state in the forbidden gap has shown in [12-15]. Topological insulator is an example for state in the forbidden band gap. The surface state of topological insulator has interesting properties that are protected by the time reversal symmetry and can not be destroyed by non-magnetic disorder. Because of these interesting properties, topological insulator is a promising material applied in nanotechnology [16-18].
In this work, in order to understand more clearly the nature of the surface plasmon polariton, we review the picture of three types of polaritons. We will analyze and compare three well-known types of polaritons to find out the similarities and differences between them. Both phonon polariton and exciton polariton have coupling constant described via longitudinal-transversal splitting. Considering a similarity of three types of polaritons as referred above, we propose a new model for bulk plasmon polariton by introducing an effective plasmon-photon coupling constant with an effective LT-splitting. Besides that, for both phonon polariton and exciton polariton there exist evanescent state in the forbidden band gap, so a question is if a similar state exists in the forbidden gap of surface plasmon polariton.

2. Phonon polariton

Phonon polariton is the result of the interaction of phonon with electromagnetic field (photon). The interaction between electromagnetic wave and phonon can be described by Maxwell equations. In the case of plane wave with the wave vector $\vec{k}$ and there is no charge in the medium, the Gauss equation is

$$\nabla\vec{D} = 0, \tag{1}$$

where $\vec{D}$ is the electric displacement vector. This equation is equivalent to

$$\varepsilon(\omega)\left(\vec{k}.\vec{E}\right) = 0, \tag{2}$$

which has two solutions $\varepsilon(\omega) = 0$ and $\vec{k}.\vec{E} = 0$. The first solution corresponds to longitudinal mode with the longitudinal resonance frequency $\omega = \omega_L$, the second solution corresponds to transversal mode with the transversal resonance frequency $\omega = \omega_T$.
Using the Maxwell equations, the dispersion relation of phonon polaritons can be expressed as follows

$$k^2 = \frac{\omega^2}{c^2}\varepsilon(\omega), \tag{3}$$

where c is velocity of photon in vacuum, and $\varepsilon(\omega)$ is the dielectric function of material,

$$\varepsilon(\omega) = \varepsilon_\infty \frac{\omega_L^2 - \omega^2}{\omega_T^2 - \omega^2}. \tag{4}$$

The solution of equation (3) is the dispersion relation of phonon polariton

$$\omega^2 = \frac{1}{2\varepsilon_\infty}\left\{c^2k^2 + \varepsilon_\infty\omega_L^2 \pm \sqrt{\left(c^2k^2 + \varepsilon_\infty\omega_L^2\right)^2 - 4\varepsilon_\infty c^2k^2\omega_T^2}\right\}. \tag{5}$$

The dispersion curve of phonon polariton is plotted in figure 1 with two branches: upper and lower. One can see that the upper branch approaches to the constant value ω_L when $k \to 0$, while the lower branch tends to $\omega = ck/\sqrt{\varepsilon_0}$. On the other hand, as $k \to \infty$ the upper branch approaches to $\omega = ck/\sqrt{\varepsilon_\infty}$ and the lower branch approaches ω_T. In the figure 2, we plot $\varepsilon(\omega)$ as a function of ω. It is easy to show that the dielectric function is positive when frequency is less than ω_T or more than ω_L, so electromagnetic wave can propagate through the material.

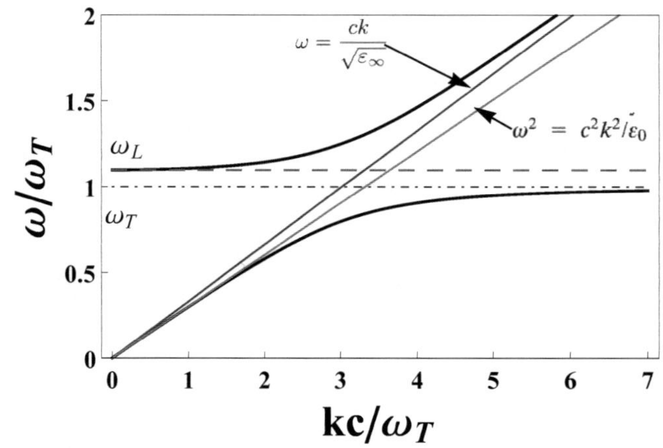

Figure 1. The dispersion curve of bulk phonon polariton in GaP (black solid line) with the values $\omega_L = 401.9cm^{-1}$, $\omega_T = 366.3cm^{-1}$, $\varepsilon_0 = 11$, $\varepsilon_\infty = 9.1$ are taken from [11].

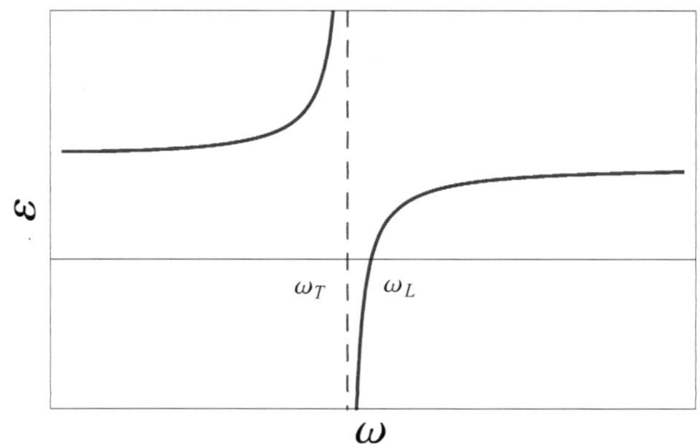

Figure 2. The dielectric function of ion crystals.

For frequencies are between ω_T and ω_L, the dielectric function is negative. In this range of frequency, electromagnetic wave does not propagate but decreases exponentially. This region is known as Reststrahlen band [19].
Phonon can interact with long-wavelength incident fields, creating a surface excitation mediated which known as surface phonon polariton. Surface phonon polariton can be excited in a polar dielectric crystal in the Reststrahlen band. When a plane wave which have frequency in the Reststrahlen band transmits to surface of a polar dielectric crystal, the positive ions move along the direction of the field and the negative ions move against it respectively. As a result the real part of permittivity is negative and an evanescent field occurs inside the polar dielectric crystal. This evanescent field is known as surface phonon polariton.
Using Maxwell equations, one can obtain the dispersion relation of surface phonon polariton as

$$\varepsilon(\omega)\sqrt{k^2 - \frac{\omega^2}{c^2}} = -\sqrt{k^2 - \varepsilon(\omega)\frac{\omega^2}{c^2}}, \tag{6}$$

therefore

$$\omega = ck\sqrt{\frac{1+\varepsilon(\omega)}{\varepsilon(\omega)}}. \tag{7}$$

We can see that the equation (6) can only be satisfied within the frequency range $\omega_T \leq \omega < \omega_L$ because the dielectric function is negative in this region. The dispersion of surface phonon polariton starts at $\omega = \omega_T$, $k = \omega_T/c$ and approaches to ω_L when $k \to \infty$ as presented in the figure 3.

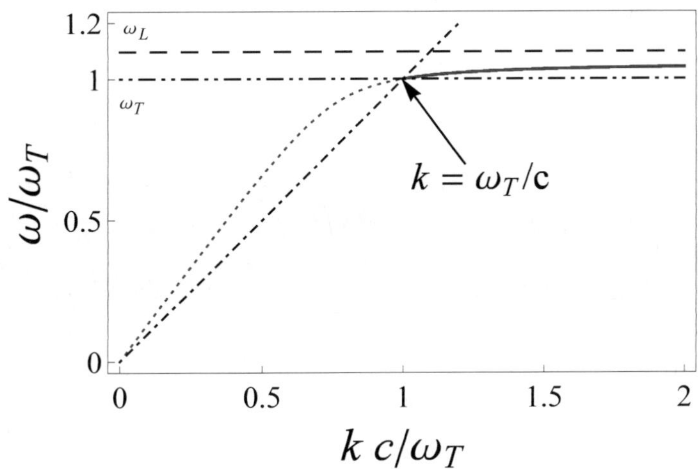

Figure 3. The dispersion curve of surface phonon polariton in GaP (blue solid line). The LT-splitting of phonon polariton is quite small, so the surface phonon polariton is neglected.

3. Exciton polariton

Exciton is an electrically neutral quasiparticle which is formed by the electrostatic Coulomb interaction between an electron and a hole. Exciton can exist in insulators, semiconductors and in some liquids. An exciton is formed when semiconductor absorbs a photon, an electron can be excited into the conduction band and a hole with positive charged is left behind in the valence band. This electron-hole pair is known as an exciton.

Exciton polariton is defined as a quasiparticle which is a result of the coupling between a photon and an exciton. The interaction of photon with exciton can be also expressed by the Maxwell equations. One can find that the dispersion relation of exciton polariton has a form like equation (5). The dispersion curve is presented in the figure 4. There are two branches which are known as upper and lower branch. For the frequency in the region between ω_T and ω_L, the electromagnetic wave decreases exponentially. The longitudinal-transversal splitting is

$$\omega_{LT} = \omega_L - \omega_T. \tag{8}$$

This splitting presents the coupling strength between the exciton and photon.

Surface exciton polariton can be referred as strong coupling of exciton with light which has a component of the wave-vector in plane of the surface. Solving the Maxwell equations with the continuity of the magnetic field at the surface, we obtain the relation between frequency and wave vector of surface exciton polariton as equations (6) and (7). One can see that, equations (6) and (7) equivalent to $\varepsilon < -1$. This condition is satisfied with frequencies in the range $\omega_T < \omega < \omega_L$. The dispersion relation of surface exciton polariton in semiconductor is presented in the figure 5.

By comparing phonon polariton and exciton polariton, we realize that both phonon polariton and

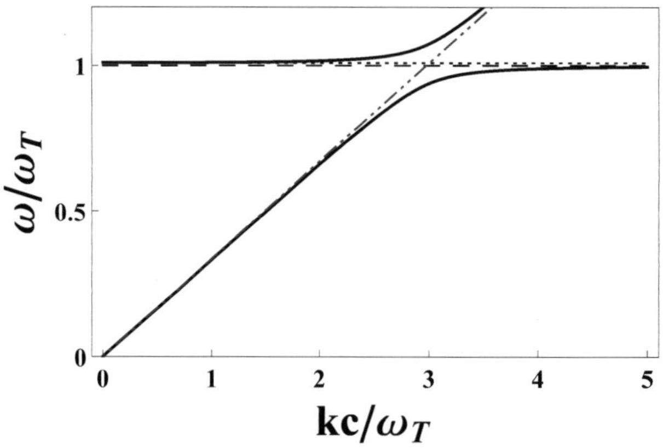

Figure 4. The dispersion of bulk exciton polariton in ZnO (black solid line).

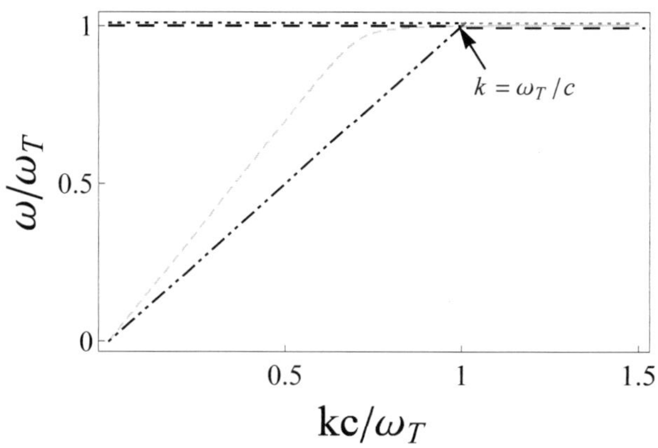

Figure 5. The dispersion of surface exciton polariton in ZnO (green solid line).

exciton polariton have longitudinal-transversal splitting that corresponds to the longitudinal-transversal symmetry breaking.
Normally, the LT-splitting of exciton polariton is very small, about meV (for GaAs $\omega_{LT} = 0.08meV$ [20], ZnO $\omega_{LT} = 5.00meV$ [21]) and the dispersion curve of surface exciton polariton is nearly straight. This leads to the mass of surface exciton polarion approaches to infinity, so it can not move and the state in this band is also neglected.

4. Plasmon polariton

Bulk plasmon polariton is quasiparticle which is a result of the coupling between plasmon and photon. This concept is similar to exciton polariton and phonon polariton. Using the Drude model for metal, we have metal dielectric constant

$$\varepsilon_m(\omega) = 1 - \frac{\omega_P^2}{\omega^2 - \omega_0^2 + i\gamma\omega}, \tag{9}$$

where γ is the collision rate, ω_0 is the frequency of restoring force, ω_P is the bulk plasma frequency, $\omega_p = \sqrt{ne^2/\varepsilon_0 m}$ in which n is the electron density, ε_0 is the vacuum permittivity and m is effective electron mass. We consider a simple case of no restoring ($\omega_0 = 0$) and small

IOP Conf. Series: Journal of Physics: Conf. Series **865** (2017) 012007 doi:10.1088/1742-6596/865/1/012007

collision rate ($\gamma = 0$), then the metal dielectric function is

$$\varepsilon_m(\omega) = 1 - \frac{\omega_P^2}{\omega^2}. \tag{10}$$

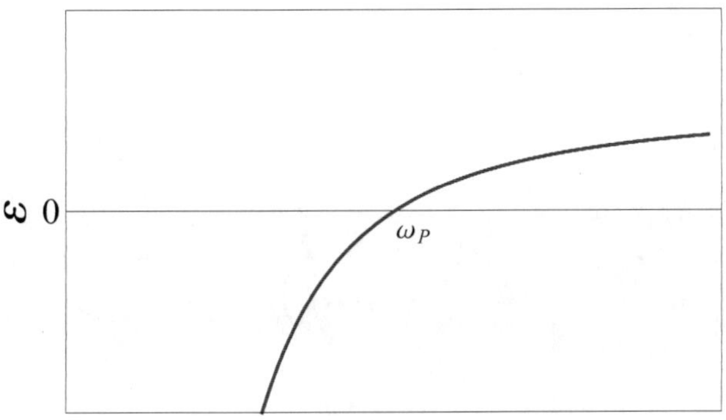

Figure 6. The dielectric function of metal (blue solid line).

Solving Maxwell equations, we obtain the dispersion relation of bulk plasmon polariton which is also represented as equation (3). The dispersion relation of bulk plasmon polariton can be rewritten in an explicit form

$$\omega^2 = \omega_P^2 + c^2k^2. \tag{11}$$

The dispersion relation implies that bulk plasmon polariton frequency is always higher than bulk plasma frequency, that means no transverse electromagnetic wave with frequency smaller than the bulk plasma frequency can travel through medium. The dispersion curve of bulk plasmon polariton is plotted in the figure 7.

Considering the similarity of three types of polaritons, we propose a new model for bulk plasmon polariton with the transversal resonance frequency is $\omega_{TBP} = 0$ and the longitudinal resonance frequency is $\omega_{LBP} = \omega_P$, then bulk plasmon polariton also has an effective bulk plasmon polariton splitting

$$\omega_{LT}^* = \omega_{LBP} - \omega_{TBP} = \omega_P. \tag{12}$$

The effective LT-splitting of bulk plasmon polariton equals to bulk plasmon energy about ten eV. This effective splitting also describes the effective coupling between bulk plasmon and photon similar to the case of phonon polariton and exciton polariton.

In the classical theory, surface plasmon polariton is electromagnetic wave bound to an interface between two media. Surface plasmon polariton can also be defined as a collective oscillation of charge carriers propagating on both side of the interface. We consider a simple model in which the dielectric medium is vacuum (the dielectric constants $\varepsilon_d = 1$). Solving the Maxwell equations at the metal/vacuum planar interface with continuity conditions of electric field, one can obtain the dispersion relation for surface plasmon as

$$k = \omega\sqrt{\frac{\varepsilon_m(\omega)}{1 + \varepsilon_m(\omega)}}. \tag{13}$$

In the Drude model for metal, the dispersion relation of surface plasmon polariton can be written in the following form

$$\omega = \sqrt{c^2k^2 + \omega_{SP}^2 - \sqrt{c^4k^4 + \omega_{SP}^4}}, \tag{14}$$

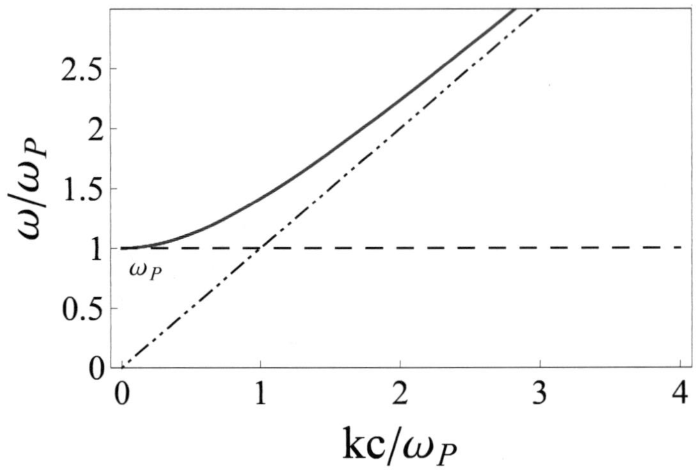

Figure 7. The dispersion curve of bulk plasmon (blue solid line).

in which $\omega_{SP} = \omega_{SP}/\sqrt{2}$. The dispersion is presented in the figure 8. This curve starts at $\omega = \omega_{TBP} = 0$ $(k = 0)$ and approaches to $\omega = \omega_{SP}$. The dispersion curve of surface plasmon polariton is in the band between ω_{TBP} and ω_{SP} similar to the case of surface phonon polariton and surface exciton polariton. There is no electromagnetic wave propagating in the band gap between ω_{SP} and $\omega_{LBP} = \omega_P$. This band gap can be called a forbidden gap.

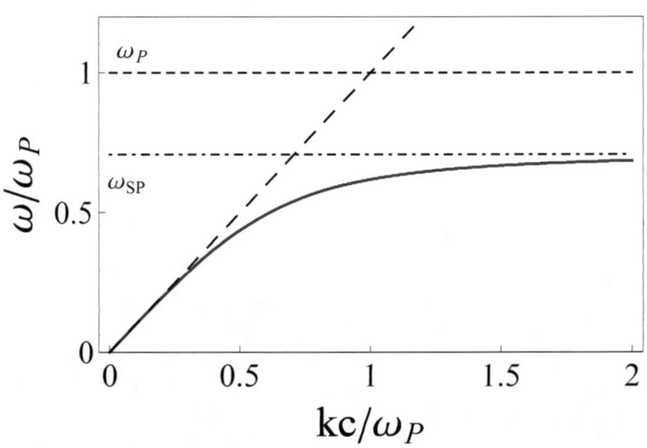

Figure 8. The dispersion curve of surface plasmon polariton (blue solid line).

In the case of phonon polariton and exciton polariton, there exists evanescent states in the band gap. The question is if a similar states in the forbidden gap of surface plasmon polariton? This problem will be investigated in our next work.

5. Conclusion

In summary, we have investigated the similarities and differences between three well-known types of polaritons: phonon polariton, exciton polariton and surface plasmon polariton.
Both phonon polariton and exciton polariton have coupling constant which is described via longitudinal-transversal splitting, while this coupling absents for the case of surface plasmon polariton because surface plasmon is longitudinal and photon is transversal polarization.

Assuming that three types of polaritons are similar, we have considered a new model for bulk plasmon polariton by introducing an effective plasmon-photon coupling constant which is described via effective LT-splitting.
The longitudinal-transversal splitting is equivalent to the longitudinal-transversal broken Galilean symmetry, so the Nambu broken symmetry theory and Anderson-Higgs mechanism could be applied to these types of polaritons.
The existence of surface modes in the LT energy gap for all three types is noted. We calculated and showed that the surface dispersion curves of three types of polaritons are similar, so a common theory of surface polaritons is needed.
There is a wide forbidden gap occurs between ω_{SP} and $\omega_{LBP} = \omega_P$. The possible existence of new mode in the energy gap of surface plasmon polariton will be the subject of our next work.

Acknowledgment
This research is funded by Vietnam National Foundation for Science and Technology Development (NAFOSTED) under grant number 103.01-2015.42.

References

[1] Feurer T, Nikolay S, David W, Vaughan J C, Statz E R, and Nelson K A 2003 *Annual Review of Materials Research* **37** 317-50
[2] Kamide K and Ogawa T 2011 *Phys. Rev.* B **83(16)** 165319
[3] Kamide K and Ogawa T 2010 *Phys. Rev. Lett.* **105(5)** 056401
[4] Chestnov I Y, Alodjants A P and Arakelian S M 2013 *Opt. Spectrosc* 115 - 363
[5] De Liberato S and Ciuti C 2009 *Phys. Rev. Lett.* **102(13)** 136403
[6] Lagoudakis K G, Wouters M, Richard M, Baas A, Carusotto I, Andr R, Dang L S and Deveaud-Pldran B 2008 *Nature Physics* **4(9)** 706-10
[7] Boulier T, Cancellieri E, Sangouard N D, Glorieux Q, Kavokin A V, Whittaker D M, Giacobino E and Bramati A 2016 *Phys. Rev. Lett.* **116(11)** 116402
[8] Antn C, Liew T C, Sarkar D, Martn M D, Hatzopoulos Z, Eldridge P S, Savvidis P G and Via L 2014 *Phys. Rev.* B **89(23)** 235312
[9] Zamfirescu M, Kavokin A, Gil B, Malpuech G and Kaliteevski M 2002 *Phys. Rev.* B **65(16)** 161205
[10] Tsintzos S I, Pelekanos N T, Konstantinidis G, Hatzopoulos Z and Savvidis P G 2008 *Nature.* **453(7193)** 372-5
[11] Peter Y and Cardona M 2010 Fundamentals of semiconductors (*Physics and material properties* vol 4) (Verlag Berlin Heidelberg: Springer) pp 244-292
[12] Richarda S, Drouhinb H J, Rougemaille N and Fishman G 2005 *J. App. Phys.* **97** 083706
[13] Nguyen T H, Drouhin H J, Wegrowe J E and Fishman G 2009*Phys. Rev.* B **79(16)** 165204
[14] Nguyen T H, Drouhin H J and Fishman G 2009 *Phys. Rev.* B **80(7)** 075207
[15] Rougemaille N, Drouhin H J, Richard S, Fishman G and Schmid A K 2005 *Phys. Rev.Lett.* **95(18)** 186406
[16] Zhang S 2015 *Scholarpedia* **10(7)** 30275
[17] Zhang X and Zhang S C 2012 *Proc SPIE Int Soc Opt Eng* **8373** 837309
[18] Jing W, Biao L, Haijun Z, Yong X and Zhang S C 2013 *Phys. Rev. Lett.* **111 (13)** 136801
[19] Caldwell J D, Lindsay L, Giannini V, Vurgaftman I, Reinecke T L, Maier S A and Glembocki O J 2015 *Nanophotonics* **4(1)** 4468
[20] Kaliteevski M. A, Brand S, Abram R A, Kavokin A and Dang L S 2007 *Phys. Rev.* B **75** 1-4
[21] Chen S L, Chen W M and Buyanova I A 2012 *Physica Status Solidi* B **249** 130711

IOP Conf. Series: Journal of Physics: Conf. Series **865** (2017) 012008 doi:10.1088/1742-6596/865/1/012008

Higher order non-classicalities in the Kerr-like coupler systems

Nguyen Thi Dung, Nguyen Manh An, Tran Thi Hai
Hong Duc University, 565 Quang Trung Rd. Dong Ve Ward, Thanh Hoa City, Vietnam
E-mail: nguyenthidungvatly@hdu.edu.vn

Abstract. In this paper, we show the existence of various types of non-classical effects in the model of the nonlinear Kerr-coupler containing two quantum nonlinear oscilators mutually coupled by continuous nonlinear interaction, such as squeezing, anti-bunching, intermodal entanglement and their higher-order counterparts. By using unitary evolution operator formalism and the common inequalities expressed in term of various moments defined by products of creation and annihilation operators, we find numerically the "exact" solutions of these factors for non-dissipative. We show and discuss the parameters considered can be indicators of generation such nonclassical effects and hence, quantumness of the system.

1. Introduction

One of the key problems related to the quantum information theory is quantum state engineering which allows us generate the nonclassical states. These special features of the quantum systems bring the remarkable applications in quantum information theory in comparison with the classical one [1, 2, 3, 4, 5, 6]. The signature of non-classicalities can be detected in the numerous ways, which can characterize as squeezing, anti-bunching, steering, intermodal entanglement, and their higher-order counter parts. Squeezed states can be suggested for the implementation of continuous variable quantum cryptography and teleportation of coherent states [1, 2]. Anti-bunching is understood as the characteristic of a radiation field in whatever place the mean number of photon is greater than the variance of the photon number distribution. Thus, antibunched states can be employed to build a high-quality single photon sources [4]. Entangled states play a crucial role in implementation quantum cryptography, quantum teleportation, dense-coding, quantum key distribution [1, 3, 5]. The existence of those properties have been widely studied in various quantum optical systems [7, 8]. However, the researchers working in different aspects of quantum information theory and quantum optics are still interested in study the possibility of generation of different nonclassical characters. In this paper we shall show how to model quantum dynamics of nonlinear coupler system, nd time-evolution wave function and evolution of operator formalism and the common inequalities expressed in term of various moments defined by products of creation and annihilation operators. Then we will show that our system exhibits its non-classical effect and hence, quantumness of the system.

2. Model of Hamiltonian

In this paper we shall consider the model comprising two mutually nonlinear interacting quantum nonlinear oscillators labeled by a and b. They are described by the following Hamiltonian expressed in terms of boson creation and annihilation operators $\hat{a}^\dagger$ ($\hat{b}^\dagger$ and $\hat{a}(\hat{b})$ respectively:

$$\hat{H} = \hbar\chi_a\hat{a}^{\dagger 2}\hat{a}^2 + \hbar\chi_b\hat{b}^{\dagger 2}\hat{b}^2 + \hbar\varepsilon_{nl}\hat{a}^{\dagger 2}\hat{b}^2 + \hbar\varepsilon^*_{nl}\hat{b}^{\dagger 2}\hat{a}^2 + \hbar\tilde{\chi}\hat{a}^\dagger\hat{a}\hat{b}^\dagger\hat{b} \tag{1}$$

The parameters χ_a (χ_b) and $\tilde{\chi}$ correspond to the self- and cross-Kerr nonlinear interaction, whereas ϵ_{nl} is a strength of the nonlinear internal coupling between two oscillators. Moreover, we assuming here that two Kerr-like oscillators located inside the cavity are identical $\chi_a/2 = \chi_b/2 = \chi$.
Thank to *Quantum Mechanics*, the quantum problem can be solved in Schrödinger picture, thus the time-evolution of the system can be understood as the evolution of the vector in Hilbert space. For state vector $|\psi(t)\rangle$, it is held by Schrödinger equation

$$i\hbar\frac{\partial}{\partial t}|\psi(t)\rangle = \hat{H}|\psi(t)\rangle \tag{2}$$

The evolution of state vector now can be determined as

$$|\psi(t)\rangle = \hat{U}|\psi(0)\rangle \tag{3}$$

where $\hat{U} = \exp\Big(-\frac{i}{\hbar}\hat{H}t\Big)$ is the unitary operator.
Using n-photon basis, we define the annihilation and creation operators of each mode as square matrices in the Hilbert space $\mathcal{H} = \mathcal{H}_a \otimes \mathcal{H}_b$ as [10]

$$\breve{a} = \hat{a} \otimes \hat{I}_n \text{ and } \breve{a}^\dagger = \hat{a}^\dagger \otimes \hat{I}_n \tag{4}$$

$$\breve{b} = \hat{I}_m \otimes \hat{b} \text{ and } \breve{b}^\dagger = \hat{I}_m \otimes \hat{b}^\dagger \tag{5}$$

for the mode a and mode b, respectively. Here $\hat{I}_{m(n)}$ is an unity matrix with $m(n)$ dimension. Applying such operators we are able to construct the Hamiltonians $\hat{H}$ and the unitary operators $\hat{U}$ in the matrix representation. Setting initial condition is $|\alpha\rangle|\beta\rangle$ in two modes of the field. These initial states can be expressed in number state as [11]

$$|\alpha\rangle = \exp\big(-\frac{|\alpha|^2}{2}\big)\sum_{n_a=0}^{\infty}\frac{\alpha^{n_a}}{\sqrt{n_a!}}|n_a\rangle; \quad |\beta\rangle = \exp\big(-\frac{|\beta|^2}{2}\big)\sum_{n_b=0}^{\infty}\frac{\beta^{n_b}}{\sqrt{n_b!}}|n_b\rangle \tag{6}$$

Thus, we can express these states in the matrix presentation. Applying standard numerical procedure we can perform and manipulate evolution of the wave-functions of the system and then calculate indicators of quantum behavior of the system expressed in terms of expectation values of functions of annihilation and creation operators of various modes.

3. Non-classicality criteria

3.1. Squeezing and higher-order squeezing

Squeezed states of electromagnetic field are defined on the basis of the generalized Heisenberg's uncertainty principle [11]. Contrary to the coherent states, they exhibit strong quantum properties. Therefore squeezing effect can be applied as indicators of the quantumness of the system. For the cases of the *one mode squeezing*, the fluctuations of two quadrature operators of a single mode a are given by the variances $\langle(\triangle\hat{X}_a)^2\rangle = \langle\hat{X}_a^2\rangle - \langle\hat{X}_a\rangle^2$ and $\langle(\triangle\hat{Y}_a)^2\rangle = \langle\hat{X}_a^2\rangle - \langle\hat{Y}_a\rangle^2$, with $\hat{X}_a$ and $\hat{Y}_a$ defined two quadrature operators [11]:

$$\hat{X}_a = \frac{\hat{a}+\hat{a}^\dagger}{2}; \quad \hat{Y}_a = -i\frac{\hat{a}-\hat{a}^\dagger}{2} \tag{7}$$

IOP Conf. Series: Journal of Physics: Conf. Series **865** (2017) 012008 doi:10.1088/1742-6596/865/1/012008

The system is squeezed if the fluctuation satisfies one of two of the follow conditions

$$4\langle(\triangle\hat{X}_a)^2\rangle - 1 < 0, \quad 4\langle(\triangle\hat{Y}_a)^2\rangle - 1 < 0. \tag{8}$$

If we express appeared above variances in terms of the creation and annihilation operators, the conditions can be written as [9]:

$$S_a = \frac{1}{2}[\langle\triangle\hat{a}^\dagger\triangle\hat{a}\rangle + \mathrm{Re}\langle(\triangle\hat{a})^2\rangle] < 0, \tag{9}$$

$$S'_a = \frac{1}{2}[\langle\triangle\hat{a}^\dagger\triangle\hat{a}\rangle - \mathrm{Re}\langle(\triangle\hat{a})^2\rangle] < 0. \tag{10}$$

More general definitions of the squeezing were defined in [9, 13], we can see as *principal squeezing.* It can be measured by the following factor

$$\lambda_a = \langle\triangle\hat{a}^\dagger\triangle\hat{a}\rangle - |\langle(\triangle\hat{a})^2\rangle| < 0 \tag{11}$$

If the fluctuation $\langle(\triangle\hat{a})^2\rangle$ is real, then two criteria are equivalent.

Two mode squeezing Analogously to the single-mode case we can define the quadrature parameters when two modes a and b are considered [14]:

$$X_{ab} = \frac{\hat{a} + \hat{a}^\dagger + \hat{b} + \hat{b}^\dagger}{2\sqrt{2}}, \quad Y_{ab} = \frac{\hat{a} - \hat{a}^\dagger + \hat{b} - \hat{b}^\dagger}{i2\sqrt{2}} \tag{12}$$

As it was shown in [15], two-mode squeezing appears when the following conditions are fulfilled

$$S_{ab} = 2[1 + \langle\triangle\hat{a}^\dagger\triangle a\rangle + \langle\triangle b^\dagger\triangle b\rangle$$
$$+2\mathrm{Re}\langle\triangle\hat{a}^\dagger\triangle b\rangle + \mathrm{Re}(\langle(\triangle a)^2\rangle + \langle(\triangle b)^2\rangle + 2\langle\triangle a\triangle b\rangle)] < 2. \tag{13}$$

$$S'_{ab} = 2[1 + \langle\triangle\hat{a}^\dagger\triangle a\rangle + \langle\triangle b^\dagger\triangle b\rangle$$
$$+2\mathrm{Re}\langle\triangle\hat{a}^\dagger\triangle b\rangle - \mathrm{Re}(\langle(\triangle a)^2\rangle + \langle(\triangle b)^2\rangle + 2\langle\triangle a\triangle b\rangle)] < 2. \tag{14}$$

Moreover, we can apply the analogous condition for the two-mode *principal squeezing*

$$\lambda_{ab} = 2[1 + \langle\triangle\hat{a}^\dagger\triangle a\rangle + \langle\triangle b^\dagger\triangle b\rangle$$
$$+2\mathrm{Re}\langle\triangle\hat{a}^\dagger\triangle b\rangle - |\langle(\triangle a)^2\rangle + \langle(\triangle b)^2\rangle + 2\langle\triangle a\triangle b\rangle|] < 2 \tag{15}$$

where

$$\langle\triangle\hat{a}^\dagger\triangle b\rangle = \langle\hat{a}^\dagger b\rangle - \langle\hat{a}^\dagger\rangle\langle b\rangle; \quad \langle\triangle a\triangle b\rangle = \langle ab\rangle - \langle a\rangle\langle b\rangle \tag{16}$$

One- and two-mode squeezing can be treated as the lowest order non-classicality indicators, whereas there appear other criteria which are can be applied to test *higher-order squeezing* effects. The concept of k-th order squeezing and its criterion was introduced for the first time by Hong and Mandel [16, 17]. However, in this part we shall concentrate on the definition proposed by Hillery [18], where higher-order squeezing is defined with use of the variances of the quadrature operators defined with use of squared creation and annihilation operators. This idea can be extended to the cases when we take into account higher powers, and two amplitude powered quadrature variables can be defined as

$$X_{h,a} = \frac{a^k + \hat{a}^{\dagger k}}{2}, \quad Y_{h,a} = -i\frac{a^k - \hat{a}^{\dagger k}}{2}. \tag{17}$$

IOP Conf. Series: Journal of Physics: Conf. Series **865** (2017) 012008 doi:10.1088/1742-6596/865/1/012008

Two operators $X_{h,a}$ and $Y_{h,a}$ do not commute and therefore, from the uncertainty relation the criterion for kth-order squeezing can be written as:

$$H_{1,a} = \langle(\triangle X_{h,a})^2\rangle - \frac{1}{2}|\langle C\rangle| < 0, \tag{18}$$

$$H_{2,a} = \langle(\triangle Y_{h,a})^2\rangle - \frac{1}{2}|\langle C\rangle| < 0, \tag{19}$$

where $[X_{h,a}, Y_{h,a}] = iC$

3.2. Normal and higher-order anti-bunching

Anti-bunching is understood as the characteristic of a radiation field in whatever place the mean number of photon is greater than the variance of the photon number distribution. In quantum statistics, for the single-mode case photon anti-bunching effect can be described in terms of the normalized second-order correlation function [19], later defined as

$$\mathcal{D}_a^2 = \langle\hat{a}^{\dagger 2}(t)a^2(t)\rangle - \langle\hat{a}^{\dagger}(t)a(t)\rangle^2 < 0. \tag{20}$$

For the case when two modes of the field a and b are considered, one can generalize two-mode anti-bunching concept and define as

$$\mathcal{D}_{ab} = \langle(\hat{a}^{\dagger}a)^2(b^{\dagger}b)^2\rangle - \langle\hat{a}^{\dagger}a\rangle^2\langle b^{\dagger}b\rangle^2 < 0. \tag{21}$$

The inequalities shown here are anti-bunching criteria determining two-mode anti-bunching, for the situation when two modes are involved, one can also apply single-mode criteria for the particular modes. Beside usual (single- and two-mode) anti-bunching effects one can consider more general case of *higher-order anti-bunching* (HOA). Single mode HOA criteria were proposed by Lee [22, 23]. After generalization, kth order anti-bunching criterion can be expressed in terms of the creation and annihilation operators as [26]:

$$\mathcal{D}_a^k = \langle\hat{a}^{\dagger k}\hat{a}^k\rangle - \langle\hat{a}^{\dagger}\hat{a}\rangle^k < 0. \tag{22}$$

From (22), it is clear that if a state is anti-bunched in kth order, it has to be anti-bunched in $(k-1)$th order, as well.

3.3. Intermodal entanglement

It is highly desirable to find inseparability criteria which would be directly applicable for multimode problems. The examples of them are those proposed by Hillery and Zubairy (HZ) [27, 28], and by Duan *at al.* [29]. The HZ inseparability criteria refer to some expectation values, variances and higher-order moments of observable operators. The first HZ criterion can be expressed in terms of the creation $\hat{a}^{\dagger}$ ($\hat{b}^{\dagger}$) and annihilation operators $\hat{a}$ ($\hat{b}$) in the following way:

$$\langle\hat{a}^{\dagger}ab^{\dagger}b\rangle - |\langle ab^{\dagger}\rangle|^2 < 0, \tag{23}$$

When this inequality is fulfilled, the state is entangled. This criterion can be generalized to the cases when higher moments are considered. For such a case it can be written as [27]

$$E_{ab}^{kl} = \langle(\hat{a}^{\dagger})^k a^k(b^{\dagger})^l b^l\rangle - |\langle a^k(b^{\dagger})^l\rangle|^2 < 0. \tag{24}$$

where k and l are non-zero positive integers.

The second HZ criterion, which is fulfilled for the separability states, can be expressed as:

$$|\langle\hat{a}\hat{b}\rangle| \leq [\langle\hat{a}^{\dagger}\hat{a}\rangle\langle\hat{b}^{\dagger}\hat{b}\rangle]^{1/2}, \tag{25}$$

and a violation of it implies that the considered state is entangled. The same way as previously, this criterion can be generalized for higher-order moments. Thus, it reads [27]:

$$E_{ab}^{\prime kl} = \langle\hat{a}^{\dagger k}a^k\rangle\langle b^{\dagger l}b^l\rangle - |\langle a^k b^l\rangle|^2 < 0 \tag{26}$$

4. Witnesses of non-classicality

4.1. Single-mode squeezing

From the solutions found in the previous section we can determine the dynamics of our quantum system. In consequence we are able to find various witnesses of non-classical phenomena. We start here from discussion of the one-mode squeezing. In Fig.1 and Fig. 2 we show the time-evolution of the squeezing parameters $S_{a(b)}$ (solid curve), $S_{a(b)}$ (dashed curve) and that of principal squeezing $\lambda_{a(b)}$. Assuming that the amplitudes α and β for the initial coherent states are real and equal each other. Because of the equivalence, the lines for two modes are identical. From the behavior of squeezing factors and principle squeezing, we see that the our system can give single mode squeezing in a and b. However, for the chosen values of the parameters, squeezing can not be created in $S_a'(b)$ factors.

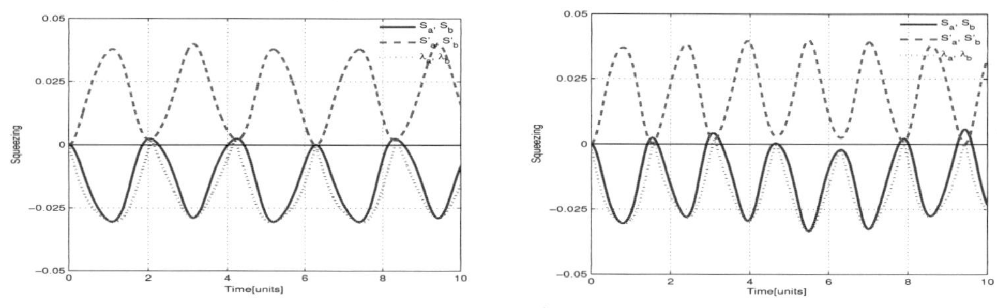

Figure 1. Evolution of squeezing factor when initial coherent state are $\alpha = 0.2$, $\beta = 0.2$. Figures a) and b) correspond to the parameters $\chi = 1$, $\tilde{\chi} = 1$, $\epsilon_{nl} = 0.5$ and $\chi = 1$, $\tilde{\chi} = 1$, $\epsilon_{nl} = 1$, respectively.

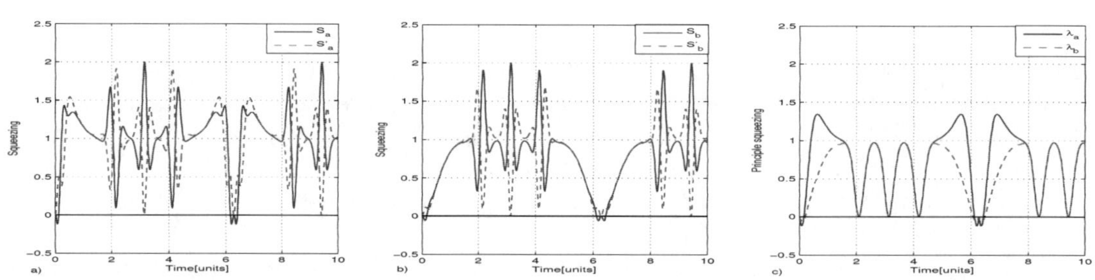

Figure 2. The time-evolution of the squeezing parameters: a) S_a, S_a' b) S_b, S_b', and c) λ_a, λ_b. Initial coherent states are $\alpha = 2$, $\beta = 0$. Other parameters: $\chi = 1, \tilde{\chi} = 1$ and $\epsilon_{nl} = 0.5$.

4.2. Two-mode squeezing

In Fig.3 two-mode quadrature variances S_{ab} (solid curve), S_{ab}' (dash curve) and two-mode principle squeezing parameter λ_{ab} are depicted. For the initial coherent states defined by $\alpha = 0.2$, $\beta = 0.2$ (see Fig. 3b)), we see that the squeezing of the quadrature S_{ab}' does not appear, contrary to that corresponding to S_{ab} which appear with a quite high intensity. Additionally, two-mode squeezing is present practically all the time in our system. When we change the initial states and assume that $\alpha = 2$ and $\beta = 0$ (see Fig. 3a), non-classical effect disappears almost completely. The most interesting point is that S'_{ab} does not give any signature of squeezing.

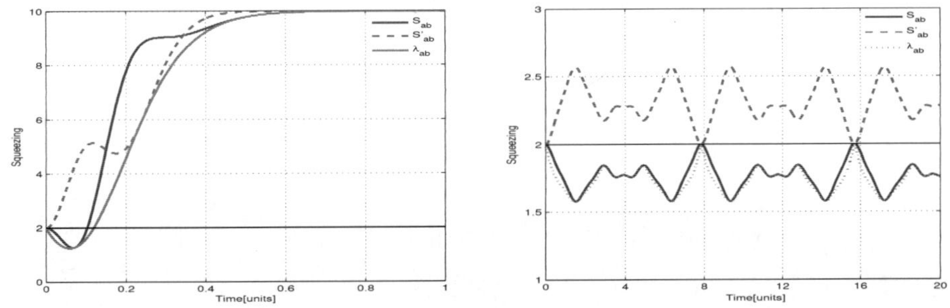

Figure 3. The time-evolution of two-mode quadrature variances: a) Initial coherent states are $\alpha = 2$, $\beta = 0$, other parameters: $\chi = 1, \tilde{\chi} = 1$ and $\epsilon_{nl} = 0.5$. b) $\alpha = 0.2$, $\beta = 0.2$, other parameters $\chi = 1, \tilde{\chi} = 1.6$ and $\epsilon_{nl} = 0.2$.

4.3. Higher order squeezing

In the preceding section, we have found the single-modes and coupled-mode squeezing effects existing in our system. In this part, we seek for the possibility of providing higher-order squeezing effect based on Hillery criteria [16, 17]. If we assume $k = 2$, the second order squeezing criteria are constructed.

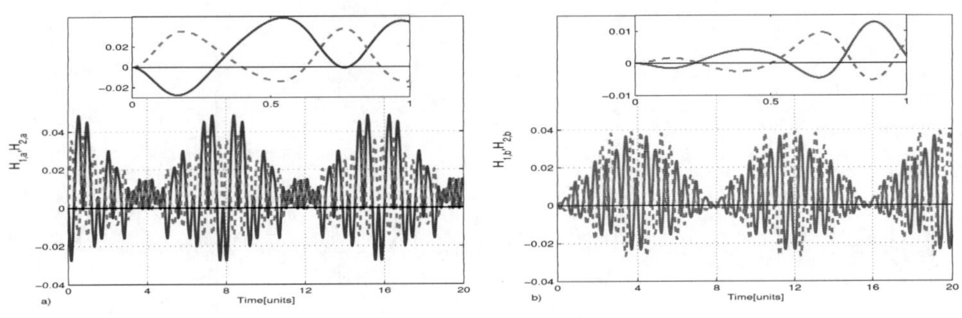

Figure 4. The time evolution of second order squeezing for mode a (Fig. a) and mode b (Fig. b) where $\alpha = 0.5$ $\beta = 0.3$, $\chi = 1, \tilde{\chi} = 1.6$ and $\epsilon_{nl} = 0.2$.

In Fig.4 and Fig. 5 we plot higher-order squeezing factor in short an long time evolution and see the negative regions of these two plots clearly illustrate the existence of this effect. Similar to the single mode and two mode squeezing, we show that the system can provide higher order squeezing and the appearance of effect in a particular quadrature depends on the choice of input coherent state.

4.4. Normal and higher-order anti-bunching effects

In this part, we proceed with the calculations of normal and higher-order anti-bunching. When the coherent single modes are equal ($\alpha = \beta$), there does not exist any signature of normal and

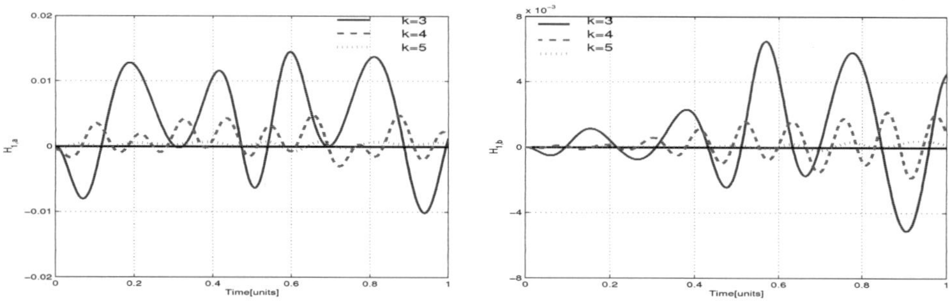

Figure 5. Short time evolution of higher-order squeezing where $\alpha = 0.5$ $\beta = 0.3$, $\chi = 1$, $\tilde{\chi} = 1.6$ and $\epsilon_{nl} = 0.2$

higher-order anti-bunching. If two initial coherent states are not equal, these effects might be pronounced.

Existence of normal anti-bunching is obtained in single-mode a for $\alpha = 2$, $\beta = 0$ and $\alpha = 0.5$, $\beta = 0.3$ is demonstrated in figures 6a) and 6b). It is easy to recognize that for our system this non-classical property is evident when the initial states are setting up with smaller values of mean number of photons. In figure 7, the results of our calculations in the factors $\mathcal{D}_a(k)$ when $\alpha = 0.5$, $\beta = 0.3$ (Fig.7a) and $\alpha = 0.5$, $\beta = 0.2$ (Fig.7 b) with $\tilde{\chi} = 1.6$, $\chi = 1$, $\epsilon_{nl} = 0.2$ for different values of k are shown. It is seen that we can only detect signals of higher-order anti-bunching in the mode a (note that the initial coherent state of mode a has greater mean number of photons than for the mode b *i.e.* $|\alpha| > |\beta|$). Otherwise, there is no any signature of higher-order anti-bunching.

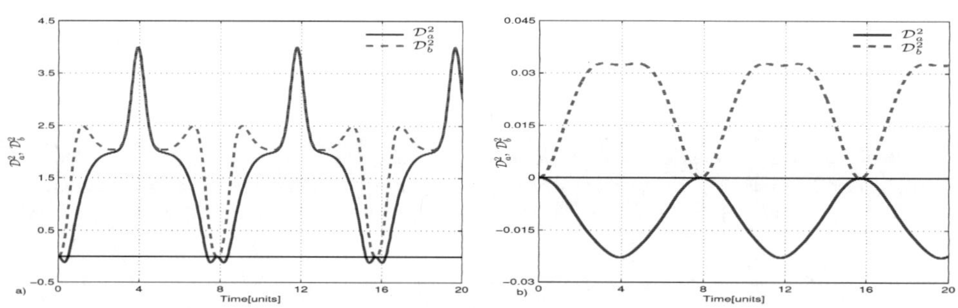

Figure 6. Normal order anti-bunching for the parameters a) $\tilde{\chi} = 1.6$, $\chi = 1$, $\epsilon_{nl} = 0.2$, $\alpha = 2$, $\beta = 0$, b) $\tilde{\chi} = 1.6$, $\chi = 1$, $\epsilon_{nl} = 0.2$, $\alpha = 0.5$, $\beta = 0.3$

4.5. Intermodal entanglement

At this point we apply two higher-order intermodal entanglement criteria for two-mode state which were introduced by Hillery and Zubairy in [27, 28].

Thus, in Fig.8 we plot $E_{ab}^{1,1}$, $E_{ab}^{2,1}$, and $E_{ab}^{2,2}$ for several different values of the nonlinear interaction strength. It is seen that two modes are not always entangled in time domain. In addition, when we increase the nonlinear interaction between two modes, we obtain stronger intermodal entanglement (see dashed lines with deeper minima). Moreover, although we observe a week

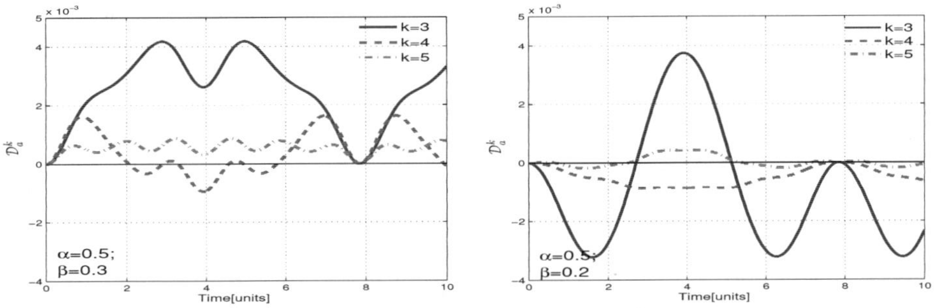

Figure 7. Evolution of higher-order anti-bunching for the parameters $\tilde{\chi} = 1.6$, $\chi = 1$, $\epsilon_{nl} = 0.2$

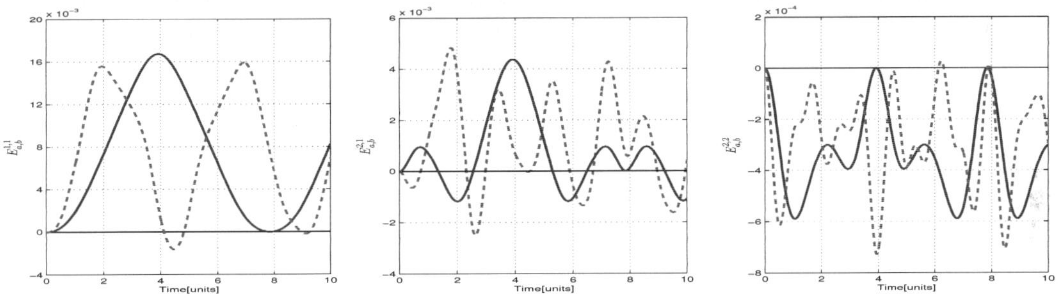

Figure 8. *Higher order intermodal entanglement for initial coherent states with* $\alpha = 0.5$, $\beta = 0.3$, *other parameters* $\tilde{\chi} = 1.6$, $\chi = 1$, $\epsilon_{nl} = 0.2$ *(solid line) and* $\epsilon_{nl} = 0.5$ *(dashed line).*

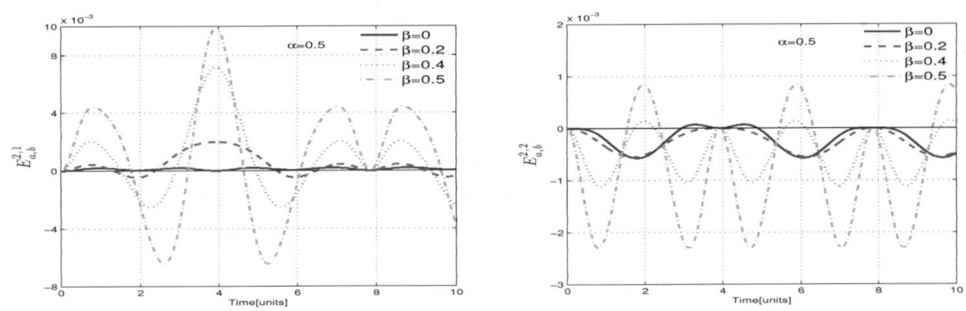

Figure 9. *Higher order intermodal entanglement for input parameters* $\tilde{\chi} = 1.6$, $\chi = 1$, $\epsilon_{nl} = 0.2$, *initial coherent state with* $\alpha = 0.5$, *and* $\beta = 0$ *(solid line),* $\beta = 0.2$ *(dashed line),* $\beta = 0.4$ *(dotted line),* $\beta = 0.5$ *(dashed-dotted line).*

signature of normal intermodal entanglement with lower value of the nonlinear interaction strength, higher-order entanglement is more pronounced (see evolution of $E_{ab}^{2,1}$, and $E_{ab}^{2,2}$). Especially, if we look at the plot corresponding to $E_{ab}^{2,2}$, we see that it is negative during the almost whole time-evolution.

In figure 9, we draw the plots of $E_{ab}^{2,1}$ ($k = 2,\ l = 1$), and $E_{ab}^{2,2}$ ($k = 2,\ l = 2$) where $\tilde{\chi} = 1.6$,

$\chi = 1$, $\epsilon_{nl} = 0.2$ for various cases of the initial coherent states. From the negative regions of the plots, we detect the signature of higher-order entanglement existing in our the system. It is seen that all plots exhibit periodic behavior. One can note that for the chosen here values of the parameters the signature of the lowest order intermodal entanglement (E_{ab}^{11}) is not detected. However, our Kerr-like coupler nonlinear system with nonlinear interaction term still can be seen as a source of intermodal entanglement and other higher-order ones in general.

5. Conclusions

Various types of non-classicality indicators of two-mode system, such as: squeezing, anti-bunching, intermodal entanglement and their higher-order counterparts were discussed. We derived the exact numerical solutions for witnesses of nonclassicality. We have analyzed the revealed effects caused by contributions of different input parameters. It is showed that the parameters considered here can be an indicators of the generation such non-classical effects and hence, quantumness of the system. The nonclassical properties of the system were more evident when we assumed small values of mean number of photons for the initial coherent states.

References

[1] Hillery M 2000 *Phys. Rev. A* **61** 022309
[2] Furusawa A, Sorensen J L, Braunstein S L, Fuchs C A, Kimble H J and Polzik E S 1998, *Science* **282** 706
[3] Bennett C H and Wiesner S J 1992 *Phys. Rev. Lett.* **69** 2881
[4] Yuan Z, Kardynal B E, Stevenson R M, Shields A J, Lobo C J, Cooper K, Beattie N S, Ritchie D A and Pepper M 2002 *Science* **295** 102
[5] Ekert A 1991 *Phys. Rev. Lett.* **67** 661
[6] Bennett C H, Brassard G, Crépeau C, Jozsa R, Peres A, and Wootters W K 1993 *Phys. Rev. Lett.* **70** 1895
[7] Sen B, Giri S K, Mandal S, Ooi C H R and Pathak A 2013 *Phys. Rev. A* **022325**
[8] Giri S K, Sen B, Ooi C H R, and Pathak A 2014 *Phys. Rev. A* **89** 033628
[9] Luks A, Perinova V and Peřina J 1988 *Opt. Commun.* **67** 149–151
[10] Nguyen T D, Leoński W, Cao Long V 2013 *CMST* **19** (3) 175–181
[11] Gerry C and Knight P 2004 *Cambridge university press*
[12] Korolkova N and Peřina J 1997 *J. Mod. Opt.* **44** 1525
[13] Tanaś R, Miranowicz A and Kielich S 1991 *Phys. Rev. A* **43** 4014
[14] Loudon R, Knight P L 1988 *J. Mod. Opt* **34** 709–759 .
[15] Kárská M and Peřina J 1990 *J. Mod. Opt.* **37**195
[16] Hong C K, Mandel L 1985 *Phys. Rev. Lett.* **54** 323:325
[17] Hong C K, Mandel L. 1985 *Phys. Rev. A* **32** 974
[18] Hillery M 1987 *Phys. Rev. A* **36** 3796
[19] Mandel L and Wolf E 1995 *Cambridge, New York*
[20] Zou X T, Mandel L 1990 *Phys. Rev. A* **41** 475
[21] Liu X M 2001 *Phys. Lett. A* **279** 123-132 .
[22] Lee C T 1990 *Phys. Rev. A.* **41** 1569
[23] Lee C 1990 *Phys. Rev. A* **41** 1721
[24] Lee C T 1990 *Phys. Rev. A.* **41**(3) 1608–1616
[25] Loudon R 1983 *Clarendon Press, Oxford* 2nd edition
[26] Gupta P, Pandey P N and Pathak A 2006 *J. Phys. B. At. Mol. Opt. Phys.* **39** 1137
[27] Hillery M and Zubairy M S 2006 *Phys. Rev. Lett.* **96** 050503
[28] Hillery M and Zubairy M S 2006 *Phys. Rev. A* **74** 032333
[29] Duan L M, Giedke G, Cirac J I and Zoller P 2000 *Phys. Rev. Lett.* **84** 2722–2725

IOP Conf. Series: Journal of Physics: Conf. Series **865** (2017) 012009 doi:10.1088/1742-6596/865/1/012009

Influence of phonon confinement on the optically detected magneto-phonon resonance line-width in parabolic quantum wells

Nguyen Dinh Hien[1], Le Dinh[1], Vo Thanh Lam[2], Tran Cong Phong[3]

[1] Center for Theoretical & Computational Physics, Hue University's College of Education, No. 34, Le Loi Str., Hue City, Viet Nam.

[2] Sai Gon University, 273 An Duong Vuong Str., Ho Chi Minh, Viet Nam.

[3] Viet Nam Institute of Educational Sciences, 101 Tran Hung Dao Str., Ha Noi, Viet Nam.

E-mail: dinhhienphysics@gmail.com

Abstract. We investigate the influence of phonon confinement described by the slab mode model on the optically detected magneto-phonon resonance (ODMPR) effect and ODMPR line-width in parabolic quantum wells by using the operator projection. The ODMPR conditions as functions of the well's parameters and the photon energy are also obtained. The shifts of ODMPR peaks caused by the confined phonon are discussed. The numerical result for a specific quantum well shows that in the two cases of confined and bulk phonons, the line-width (LW) is sensitive to the change of the confinement frequency, and temperature. Furthermore, the influence of slab mode phonon plays an important role and cannot be neglected in reaching the ODMPR line-width for all range of the well's confinement frequency.

1. Introduction

Magneto-phonon resonance (MPR) effect is known to arise from the resonant coupling between electrons and optical phonons when the separation between two Landau levels approaches the phonon energy. The absorption line-width is well-known as a good tool for investigating the scattering mechanisms of carriers. Hence, it can be used to probe electron-phonon scattering processes. The line-width is defined as the profile of curves describing the dependence of the absorption power on the photon energy or frequency [1]. The line-width has been measured in various kinds of semiconductors, such as quantum wells [2], quantum wires [3], and quantum dots [4]. These results show that the absorption line-width has a weak dependence on temperature and has a strong dependence on the sample size. However, in these studies, the absorption line-width was investigated based on the interaction of electrons and bulk phonons.

Phonon confinement is an essential part of the description of electron-phonon interactions [5]. It causes the increase of electron-phonon scattering rates [6, 7] and modifies the phonon density of states [8]. Phonon confinement affects magneto-phonon resonance mainly through changes in the selection rules for transitions involving subbands of electrons, Landau levels, phonon modes and affects the line-width through changes in the probability of the electron-phonon scattering.

Recently, our group has proposed a method, called the profile method. This method can be used to computationally obtain the line-width from graphs of the absorption power [9]. We used this method to determine the cyclotron resonance line-width and the influence of phonon confinement on the optically detected electro-phonon resonance line-width in quantum wires [10, 11, 12], and in square quantum wells [13, 14]. However, the absorption line-width in parabolic quantum wells (PQWs) due to confined optical phonon-electron interaction has not been considered in previously published reports.

In the present work, we investigate influence of phonon confinement described by slab mode model on the ODMPR line-width (ODMPRLW) in a PQW. The dependence of the ODMPRLW on the well's confinement frequency is obtained. The results of the present work are fairly different from the previous theoretical results because the slap mode model of phonon confinement are considered in detail. The paper is organized as follows. In Section 2, we introduce slab mode model of phonon in a PQW. Derivations of analytical expression of the absorption power in PQWs due to slab LO phonon modes are presented in Section 3. The graphical dependence of the absorption power on the photon energy in the GsAs/AlAs PQW is shown in Section 4. From this dependence, we obtain the dependence of the line-width on temperature and well's confinement frequency. The results are compared with the corresponding calculations for bulk phonon model. Finally, remarks and conclusions are shown in Section 5.

2. Slab mode model of phonon confinement

We consider a single parabolic quantum well structure where magnetic field $\vec{B} = (0, 0, B)$ is applied in the z direction. Suppose that the Landau gauge of vector potential is $\vec{A} = (0, xB, 0)$, the one-electron eigenfunction, $|\alpha\rangle = \Psi_{N,n,k_y}$, is given by [15]

$$\Psi_{N,n,k_y} = \frac{1}{\sqrt{L_y}} \exp(ik_y y)\psi_N(x - x_0)\varphi_n(z), \tag{1}$$

where $N = 0, 1, 2, \cdots$ is the Landau level index; $n = 0, 1, 2, \cdots$ is the electric subband quantum number; $\psi_N(x - x_0)$ is the harmonic oscillator wave function centered at $x_0 = -a_c^2 k_y$ with $a_c = (\hbar/(m^*\omega_c))^{1/2}$ being the cyclotron radius of the orbit in the (x, y) plane; L_y and k_y are specimen dimension and electron wave-vector in y-direction; $\varphi_n(z)$ is the electron wave function in z-direction as determined by the parabolic potential $V(z)$, $V(z) = m^*\omega_z^2 z^2/2$. The corresponding energy eigenvalues, E_{Nn}, is given by [16]

$$E_{N,n} = (N + 1/2)\hbar\omega_c + (n + 1/2)\hbar\omega_z, \tag{2}$$

with $\omega_c = eB/m^*$, ω_z, and m^* are the cyclotron frequency, well's confinement frequency, and effective mass of an electron, respectively. In this case, $\varphi_n(z)$ is given by [16]

$$\varphi_n(z) = \frac{1}{\sqrt{2^n n!\sqrt{\pi}a_z}} \exp(-\frac{z^2}{2a_z^2}) H_n(\frac{z}{a_z}), \tag{3}$$

with $a_z = (\hbar/(m^*\omega_z))^{1/2}$, and H_n are Hermite polynomials of order n.

The matrix element for electron-slab mode phonon interaction in parabolic quantum wells in the extreme quantum limits can be written as [16]

$$|\langle i|H_{e-ph}|f\rangle|^2 = \frac{e^{2-}h\omega_{LO_{m,q_\perp}}\chi^*}{2\epsilon_0 V_0} |V_{m\phi}^S(q_\perp)|^2 |J_{NN'}(u)|^2 |G_{nn'}^{Sm\phi}|^2 \delta_{k'_\perp, k_\perp \pm q_\perp}, \tag{4}$$

where

$$|J_{NN'}(u)|^2 = \frac{n_2!}{n_1!} e^{-u} u^{n1-n2} [L_{n2}^{n1-n2}(u)]^2, \tag{5}$$

with $u = a_c^2 q_\perp^2/2$, $n1 = \max\{N, N'\}$, $n2 = \min\{N, N'\}$, and L_{n2}^{n1-n2} is the Laguerre polynomial. The overlap integral, $G_{nn'}^{Sm\phi}$, is given by

$$G_{nn'}^{Sm\phi} = \int_{-\infty}^{\infty} \varphi_{n'}^*(z) u_{m\phi}^S(z) \varphi_n(z) dz, \tag{6}$$

where $\chi^* = (\chi_\infty^{-1} - \chi_0^{-1})$ with χ_∞ and χ_0 are the high and static-frequency dielectric constants, respectively; ϵ_0 is the vacuum dielectric constants, and $V_0 = SL_z$ is the volume of the system; $\hbar\omega_{LO_{m,q_\perp}} = \hbar\sqrt{\omega_{LO}^2 - \beta^2(q_\perp^2 + q_m^2)}$ is the energy of confined optical phonon; $\hbar\omega_{LO}$ is the energy of bulk optical phonon; β is the velocity parameter; $q_\perp$ is a two dimensional vector in the (x, y) plane of phonon; $q_m = m\pi/L_z$; $u_{m\phi}^S(z)$ is the parallel component of the displacement vector of $m-$th phonon mode in the direction of the spatial confinement for the slab mode model (S) [17]; ϕ are the even $(-)$ and odd $(+)$ slab mode phonons. In the next section, we will use the slab mode model to calculate the optical absorption power in quantum wells. For the slab mode model, we obtain

$$u_{m+}^S(z) = \cos\left(\frac{m\pi z}{L_z}\right),\ m = 1, 3, 5, \cdots, \tag{7}$$

$$u_{m-}^S(z) = \sin\left(\frac{m\pi z}{L_z}\right),\ m = 2, 4, 6, \cdots. \tag{8}$$

3. Analytical results

We utilize slab mode model for confined phonon to calculate the absorption power in above mentioned PQWs, subjected to an ac electromagnetic field with amplitude E_0 and frequency ω. The absorption power is obtained by relating it to the transitions probability of the photon absorption to move to the higher energy levels along with phonon absorption or emission processes as follows [18]

$$P(\omega) = \frac{E_0^2}{2\hbar\omega} \sum_\alpha \frac{|j_\alpha^+|^2 (f_\alpha - f_{\alpha+1}) B(\omega)}{(\omega - \omega_c)^2 + [B(\omega)]^2}, \tag{9}$$

where $|j_\alpha^+|^2 = |\langle\alpha + 1|j^+|\alpha\rangle|^2 = (N + 1)(2e^2\hbar\omega)/m^*$; f_α and $f_{\alpha+1}$ are the Fermi-Dirac distribution functions of electron at state $|\alpha\rangle = |N, n, k_y\rangle$ and $|\alpha + 1\rangle = |N + 1, n, k_y\rangle$. The term $B(\omega)$ is given by

$$\begin{aligned} B(\omega) &= \frac{e^2\hbar\omega_{LO_{m,q_\perp}}\chi^*}{8\pi\hbar\epsilon_0 L_z} \sum_{N',n'} \sum_{m,\phi=\pm} \frac{|G_{n'n}^{Sm\phi}|^2}{(f_{N+1,n} - f_{N,n})} \int_0^\infty q_\perp dq_\perp \frac{|J_{N,N'}(u)|^2}{a_{m\phi}^S q_\perp^2 + \frac{b_{m\phi}^S}{L_z^2}} \\ &\times \{[(1 + N_{\vec{q}}) f_{N+1,n}(1 - f_{N',n'}) - N_{\vec{q}} f_{N',n'}(1 - f_{N+1,n})]\delta(E_1^-) \\ &+ [(1 + N_{\vec{q}}) f_{N',n'}(1 - f_{N+1,n}) - N_{\vec{q}} f_{N+1,n}(1 - f_{N',n'})]\delta(E_1^+)\} \\ &+ \frac{e^2\hbar\omega_{LO_{m,q_\perp}}\chi^*}{8\pi\hbar\epsilon_0 L_z} \sum_{N',n'} \sum_{m,\phi=\pm} \frac{|G_{nn'}^{Sm\phi}|^2}{(f_{N+1,n} - f_{N,n})} \int_0^\infty q_\perp dq_\perp \frac{|J_{N+1,N'}(u)|^2}{a_{m\phi}^S q_\perp^2 + \frac{b_{m\phi}^S}{L_z^2}} \\ &\times \{[(1 + N_{\vec{q}}) f_{N',n'}(1 - f_{N,n}) - N_{\vec{q}} f_{N,n}(1 - f_{N',n'})]\delta(E_2^-) \\ &+ [(1 + N_{\vec{q}}) f_{N,n}(1 - f_{N',n'}) - N_{\vec{q}} f_{N',n'}(1 - f_{N,n})]\delta(E_2^+)\}, \end{aligned} \tag{10}$$

with $N_{\vec{q}}$ is the Planck distribution function for slab mode phonon at the state $|q\rangle = |m, q_\perp\rangle$,

$$E_1^\pm = \hbar\omega + (N' - N - 1)\hbar\omega_c + (n' - n)\hbar\omega_z \pm \hbar\omega_{LO_{m,q_\perp}}, \tag{11}$$

$$E_2^\pm = \hbar\omega + (N - N')\hbar\omega_c + (n - n')\hbar\omega_z \pm \hbar\omega_{LO_{m,q_\perp}}, \tag{12}$$

the Dirac delta functions ($\delta(E_\ell^\pm)$, $\ell = 1,2$) in Eq. (10) are replaced by Lorentzians of width $\gamma_{N,N'}^{S\pm}$ and $\gamma_{N+1,N'}^{S\pm}$, namely [19]

$$\delta(E_1^\pm) = \frac{1}{\pi}\frac{\gamma_{N,N'}^{S\pm}}{(E_1^\pm)^2 + (\gamma_{N,N'}^{S\pm})^2},\ \delta(E_2^\pm) = \frac{1}{\pi}\frac{\gamma_{N+1,N'}^{S\pm}}{(E_2^\pm)^2 + (\gamma_{N+1,N'}^{S\pm})^2}, \tag{13}$$

where

$$(\gamma_{N',N}^{S\pm})^2 = \frac{e^2\hbar\omega_{LO_{m,q_\perp}}\chi^*}{8\pi^2\hbar\epsilon_0 V_0}(N_q + \frac{1}{2} \pm \frac{1}{2}) \sum_{m,\phi=\pm} |G_{n'n}^{Sm\phi}|^2 \int_0^\infty q_\perp dq_\perp \frac{J_{N,N'}}{a_{m\phi}^S q_\perp^2 + \frac{b_{m\phi}^S}{L_z^2}}, \tag{14}$$

$$(\gamma_{N+1,N'}^{S\pm})^2 = \frac{e^2\hbar\omega_{LO_{m,q_\perp}}\chi^*}{8\pi^2\hbar\epsilon_0 V_0}(N_q + \frac{1}{2} \pm \frac{1}{2}) \sum_{m,\phi=\pm} |G_{nn'}^{Sm\phi}|^2 \int_0^\infty q_\perp dq_\perp \frac{J_{N+1,N'}}{a_{m\phi}^S q_\perp^2 + \frac{b_{m\phi}^S}{L_z^2}}. \tag{15}$$

We can see that these analytical results appear very involved. However, physical conclusion can be drawn from graphical representations and numerical results, obtained from adequate computational methods.

4. Numerical results and Discussion

To clarify the obtained analytical results we numerically evaluate the absorption power, $P(\omega)$, for a specific GaAs/AlAs PQW. The absorption power is considered to be a function of the photon energy. The parameters used in our computational evaluation are as follows [20]: $\chi_\infty = 10.9, \chi_0 = 12.9, m^* = 0.067 \times m_0$ (m_0 being the mass of free electron), $E_0 = 5.0 \times 10^6$ Vm^{-1}, $\hbar\omega_{LO} = 36.250$ meV, and $\beta = 4.73 \times 10^3\ \mathrm{ms}^{-1}$. In this case, for example, we have $\hbar\omega_{LO_{1,q_\perp}} = \hbar\sqrt{\omega_{LO}^2 - \beta^2(q_\perp^2 + q_1^2)} = 36.2467\mathrm{meV} \simeq \hbar\omega_{LO}$. It means that optical phonons are actually weakly dispersive, confined phonon energies in the different levels ($\hbar\omega_{LO_{m,q_\perp}}$) differ from bulk phonon ones ($\hbar\omega_{LO}$) very little [14, 21]. The following results are obtained in the extreme quantum limit, and assuming that only the lowest electric subbands are occupied by the electrons: $n = 0, n' = 0, 1$ for confined electron and Landau levels $N = 0, N' = 1$.

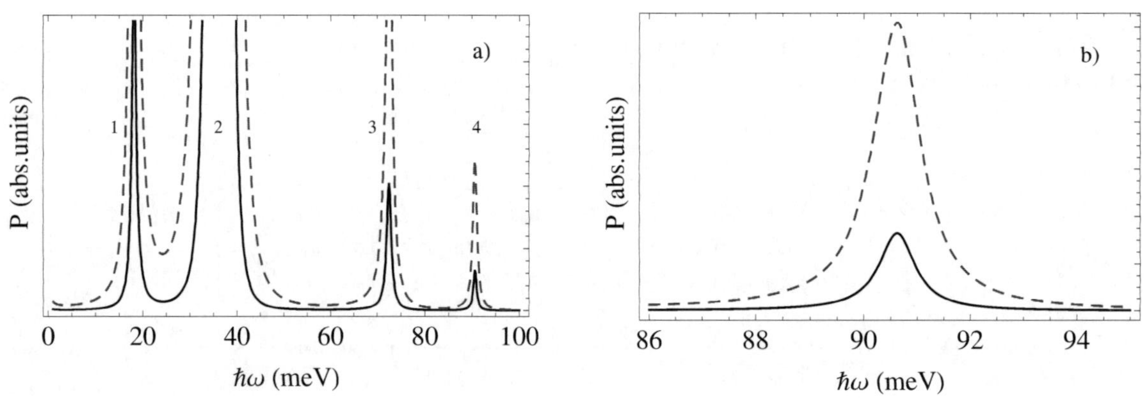

Figure 1. a) Dependence of the absorption power in PQW on the photon energy for two different models of phonon: bulk phonons (the solid curve) and confined phonons (the dashed curve). Here, $T = 300$ K, $B = 20.97$ T, and $\omega_z = 0.5\omega_{LO}$. Figure b) is at the fourth peak in Fig. a).

Figure 1a) shows the dependence of the absorption power on the photon energy for bulk phonons (solid curve), confined phonons described by the slap model (dashed curve). The

IOP Conf. Series: Journal of Physics: Conf. Series **865** (2017) 012009 doi:10.1088/1742-6596/865/1/012009

result is calculated for a parabolic quantum well, at $T = 300$ K, $B = 20.97$ T, and $\omega_z = 0.5\omega_{LO}$. There are four peaks in each curve, which correspond to the ODMPR for intra-subband transitions (0-0) and inter-subband transitions (0-1). For intra-subband transitions, the ODMPR peaks satisfy the condition $\hbar\omega = (N' - N)\hbar\omega_c \pm \hbar\omega_{LO}$, while this condition is $\hbar\omega = (N' - N)\hbar\omega_c + (n - n')\hbar\omega_z \pm \hbar\omega_{LO}$ for inter-subband transitions.

- The first peak corresponds to the photon energy $\hbar\omega = 18.125$ meV, which satisfies the condition $\hbar\omega = (N' - N)\hbar\omega_c - (n - n')\hbar\omega_z - \hbar\omega_{LO}$, i.e., 18.125 meV$= (1 - 0) \times 36.250 - (0 - 1) \times 0.5 \times 36.250$ meV $-$ 36.250 meV. This is the condition for ODMPR with $P \equiv N' - N = 1$. It describes the transitions of an electron between both the Landau levels $N = 0, N' = 1$ and the inter-subband $n = 0, n' = 1$ by absorbing a photon with energy $\hbar\omega$ along with the absorption of a phonon with energy $\hbar\omega_{LO}$.

- The second peak corresponds to the photon energy $\hbar\omega = 36.250$ meV, which satisfies the condition $\hbar\omega = (N' - N)\hbar\omega_c$. Therefore, this peak is called cyclotron resonance one with $P = 1$. It describes the transition of an electron from the initial Landau level $N = 0$ to the another Landau level $N' = 1$ by absorbing a photon with energy $\hbar\omega$.

- The third peak corresponds to the photon energy of $\hbar\omega = 72.500$ meV, which satisfies the condition $\hbar\omega = (N' - N)\hbar\omega_c + \hbar\omega_{LO}$, i.e., 72.500 meV$= (1 - 0) \times 36.250$ meV $+$ 36.250 meV. This is the condition for ODMPR with $P = 1$. This condition implies that an electron in the Landau level $N = 0$ can move to an another Landau level $N' = 1$ by absorbing a photon with energy $\hbar\omega$ along with the emission of a phonon with energy $\hbar\omega$.

- The fourth peak corresponds to the photon energy $\hbar\omega = 90.625$ meV, which satisfies the condition $\hbar\omega = (N' - N)\hbar\omega_c - (n - n')\hbar\omega_z + \hbar\omega_{LO}$, i.e., 90.625 meV$= (1 - 0) \times 36.250 - (0 - 1) \times 0.5 \times 36.250$ meV $+$ 36.250 meV. This is the condition for ODMPR with $P = 1$. It describes the transitions of an electron between both the Landau levels $N = 0, N' = 1$ and the inter-subband $n = 0, n' = 1$ by absorbing a photon with energy $\hbar\omega$ along with the emission of a phonon with energy $\hbar\omega_{LO}$.

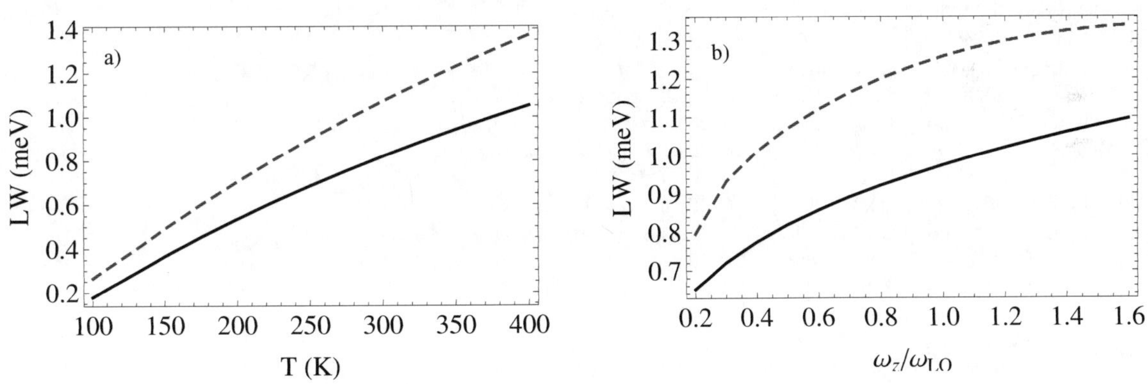

Figure 2. Dependence of the LW for two different models of phonon. The dashed curve and the solid curve correspond to the case of confined phonons and bulk phonons, respectively: a) on the temperature. Here, $B = 20.97$ T and $\omega_z = 0.5\omega_{LO}$. b) on the well's confinement frequency. Here, $T = 300$ K, and $B = 20.97$ T.

Figure 2a) shows that the LW increases with increasing temperature in both models of phonon. This is reasonable because the possibility of the electron-phonon scattering rises as the temperature increases. It is also seen that the LW of the ODMPR peak in the case of confined phonon varies faster and has a larger value than the bulk one. This is because when phonons are confined the probability electron-phonon scattering is increased.

Figure 2b) shows that the LW increases with increasing well's confinement frequency for both models of the phonon. This can be explained physically by a increases in the possibility of electron-phonon scattering when the well's confinement frequency increases. Furthermore, the LW for the confined phonon case varies faster and has a larger value than it does for the bulk phonon case. Thus, in all range of the confinement frequency, the influence of phonon confinement plays an important role and cannot be neglected in reaching the ODMPR line-width.

5. Conclusions

We have calculated analytical expressions for the absorption power in parabolic quantum wells due to confined electrons and LO slab mode-confined phonons interaction. From the graphs of the absorption power, we obtained the LW as a profile of curves. Computational results show that in the cases of both bulk and confined phonons, the LW increases with temperature and well's confinement frequency. In addition, the LW for the confined phonon case has a larger value and varies faster than it does for the bulk phonon case. Thus, in all range of the well's confinement frequency, the influence of phonon confinement plays an important role and cannot be neglected in reaching the ODMPR line-width. This result is the same as that obtained in a two-dimensional system of ref. 7 which is verified by theory and experiments.

Acknowledgments

This work was supported by MOET-Vietnam in the scope research project coded of B2014-45-01.

References

[1] Cho Y J, Choi S D 1994 *Phys. Rev.* B **49** 14301
[2] Spector H N, Lee J, Melman P 1986 *Phys. Rev.* B **34** 2554
[3] Weman H, Sirigu L, Karlsson K F, Leifer K, Rudra A, Kapon E 2002 *Appl. Phys. Lett.* **81** 2839
[4] Matthiesen C, Vamivakas A N, Atatre M 2012 *Phys. Rev. Lett.* **108** 093602
[5] Bennett C R, Guven K, Tanatar B 1998 *Phys. Rev.* B **57** 3994
[6] Nishiguchi N 1995 *Phys. Rev.* B **52** 5279
[7] Rudin S, Reinecke T L 1990 *Phys. Rev.* B **41** 7713
[8] Svizhenko A, Balandin A, Bandyopadhyay S, Stroscio M A 1998 *Phys. Rev.* B **57** 4687
[9] Phong T C, Phuc H V 2011 *Mod. Phys. Lett.* B **25** 1003
[10] Phong T C, Phuong L T T, Phuc H V 2012 *Superlattices Microstruct.* **52** 16
[11] Phuong L T T, Phuc H V, Phong T C 2014 *Physica* E **56** 102
[12] Phong T C, Phuong L T T, Phuc H V, Vinh P T 2013 *J. Korean Phys. Soc.* **62** 305
[13] Phuong L T T, Phuc H V, Phong T C 2014 *Physica* E **56** 102
[14] Phong T C, Phuong L T T, Hien N D, Lam V T 2015 *Physica E* **71** 79
[15] Sang Chil Lee et al 2005 *Physica* E **28** 402
[16] Bhat J S, Mulimani B G, Kubakaddi S S 1994 *Phys. Rev.* B **49** 16459
[17] Licari J J, Evrard R 1977 *Phys. Rev.* B **15** 2254
[18] Kang N L, et al. 1995 *J. Phys.: Condens. Matter* **7** 8629
[19] Chaubey M P, Van Vliet C M 1986 *Phys. Rev.* B **33** 5617
[20] Masale M, Constantious N C 1993 *Phys. Rev.* B **48** 11128
[21] Hien N D, Dinh L, Lam V T, Phong T C 2016 *Journal of Physics: Conference Series* **726** 012012

IOP Conf. Series: Journal of Physics: Conf. Series **865** (2017) 012010 doi:10.1088/1742-6596/865/1/012010

Grand−canonical Monte−Carlo simulation of DNA condensation in equilibrium with a salt mixture containing 2:2 salt

Nguyen V Duc[2], **Toan T Nguyen**[1,2,3]

[1]VNU Key Laboratory for Multiscale Simulation of Complex Systems, Vietnam National University, 334 Nguyen Trai Street, Thanh Xuan, Hanoi, Vietnam

[2]Faculty of Physics, VNU University of Science, Vietnam National University, 334 Nguyen Trai Street, Thanh Xuan, Hanoi, Vietnam

[3]School of Physics, Georgia Institute of Technology, 837 State Street, Atlanta, Georgia 30332-0430, USA

E-mail: ducnv84@gmail.com

Abstract. The Grand−canonical Monte−Carlo simulation method is used to simulate DNA hexagonal condensate in the presence of a mixture of 1:1, 2:1, 2:2 salts using the primitive ion model. Previous results show that DNA can be condensed by divalent counterions in restricted environment, such as inside viruses. In this work, we study the effects of divalent co-ions on condensation of DNA by divalent counterions. It is shown that divalent co−ions lead to weaker DNA condensation free energy and DNA de−condensation at smaller counterions concentration.

1. Introduction

The problem of DNA condensation in the presence of multivalent counterions has seen a strong revival of interest in recent years. This is because of the promising development of DNA-based nanotechnology that can be controlled exquisitely at nanoscale into precise 2D and 3D shapes. This has the potential to enables the fabrication of precise nanoscale devices that have already shown great potential for biomedical applications such as: drug delivery, biosensing, and synthetic nanopore formation [1, 2]. Furthermore, DNA study enables us to find effective ways of gene delivery for the rapidly growing field of genetic therapy. DNA viruses such as bacteriophages provide excellent study candidates for this purpose. One can package genomic DNA into viruses, then deliver and release the molecule into targeted individual cells. Recently there is a large biophysics literature dedicated to the problem of DNA condensation (packaging and ejection) inside bacteriophages [3].

Because DNA is a strongly charged molecule in aqueous solution, electrostatics and the screening condition of the solution play an important role in the structure and functions of DNA systems. Specifically, the condensation of DNA molecules is strongly influenced by the counterion valence [4, 5, 6, 7, 8, 9]. The strong influence of multivalent counterions on DNA ejection from viruses is expected to have the same physical origin as the phenomenon of reentrant DNA condensation in free solution [10, 11, 12, 13, 14]. The marginal case of divalent counterions can be also understood. While divalent counterions are able to condense DNA only partially in free solution [4, 5], DNA virus provides a unique experimental setup. The constraint of the viral capsid strongly eliminates configurational entropic cost of packaging DNA. This allows divalent counterions to influence DNA condensation similar to that of trivalent/tetravalent counterions. DNA condensation by divalent counterions has also been observed in another environment where DNA configuration is constrained, namely the condensation of DNA in two dimensional systems [15].

In this paper, we study the problem of DNA condensation in the presence of divalent counterions using Grand canonical Monte Carlo computer simulations within primitive ion model, expanding upon previous works from our groups [16, 17, 18]. We study numerically the osmotic pressure of DNA hexagonal bundle in equilibrium with a bulk solution of salt mixture of monovalent and divalent counterions. We give some preliminary result on the effect of divalent co−ions. The paper is organized as follows. In Sec.2, the model of our system and various physical parameters used in the simulation are presented in details. In Sec.3, the results are presented and their relevance to available experimental data is discussed. We conclude in Sec.4.

2. The simulation model and method

We assume the DNA molecules in the bundle to arrange in a two dimensional hexagonal lattice with lattice constant d (see Fig. 1). The DNA axis is parallel to the Z-axis. Individual DNA

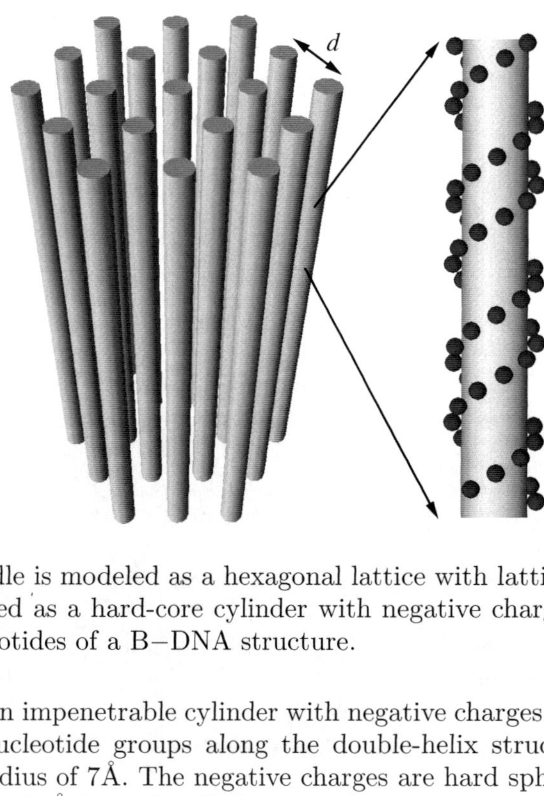

Figure 1. A DNA bundle is modeled as a hexagonal lattice with lattice constant d. Individual DNA molecule is modeled as a hard-core cylinder with negative charges glued on it according to the positions of nucleotides of a B−DNA structure.

molecule is modeled as an impenetrable cylinder with negative charges glued on it in accordance with the locations of nucleotide groups along the double-helix structure of a B−DNA. The hardcore cylinder has radius of 7Å. The negative charges are hard spheres of radius 2Å, charge $-e$ and lie at a distance of 9Å from the DNA axis. This gives an averaged DNA radius, r_{DNA}, of 1nm. The solvent water is treated as a dielectric medium with dielectric constant $\varepsilon = 78$ and temperature $T = 300^oK$. The positions of DNA molecules are fixed in space. This mimics the constraint on DNA configurational entropy inside viruses and other experiments of DNA condensation using divalent counterions in restricted environment. The mobile ions in solution are modeled as hard spheres with unscreened Coulomb interaction (the primitive ion model). For simplicity, all ions have radius of $\sigma_x = 2$Å.

IOP Conf. Series: Journal of Physics: Conf. Series **865** (2017) 012010 doi:10.1088/1742-6596/865/1/012010

The simulation is carried out using the periodic boundary condition. A periodic simulation cell with $N_{DNA} = 12$ DNA molecules in the horizontal (x, y) plane and 3 full helix periods in the z direction is used. In Ref. [19, 20], it is shown that the macroscopic limit is reached when $N_{DNA} \geq 7$. Our simulation cell contains 12 DNA helices, hence it has enough DNA molecules to eliminate the finite size effect.

The simulation is carried out using the Grand canonical ensemble where the number of ions is not constant during the simulation. Instead their chemical potentials are fixed. These chemical potentials are chosen in advance by simulating a DNA−free salt solution and adjusting them so that the solution has the correct ion concentrations. If a state i of the system that is characterized by the locations of N_{iZ+} multivalent counterions, N_{i+} monovalent counterions, N_{iZ-} multivalent counterions, N_{i-} coions ($Z = 2$ for divalent ions), in the grand canonical ensemble of unlabeled particles, the probability of such state is given by

$$\pi_i = \frac{1}{\mathcal{Z}} \frac{1}{\Lambda_{Z+}^{3N_{iZ+}} \Lambda_{+}^{3N_{i+}} \Lambda_{Z-}^{3N_{iZ-}} \Lambda_{-}^{3N_{i-}}} \exp\left[\beta(\mu_{Z+}N_{iZ+} + \mu_+ N_{i+} + \mu_- N_{i-} + \mu_{Z-}N_{iZ-} - U_i\right] \quad (1)$$

Here, $\mathcal{Z}$ is the grand canonical partition function, $\beta = 1/k_BT$, $\Lambda_x \equiv h/\sqrt{2\pi m_x k_B T}$ are the thermal wavelength of the corresponding ion type (here x are either $Z+$, $Z-$, $-$ or $+$), U_i is the interaction energy of the state i, and μ_x are the corresponding chemical potential of the ions. This formula is the basis for Grand canonical Monte−Carlo simulation. More details on this method can be found from previous works[18, 17, 21]. Unless explicitly stated otherwise, through this work, we simulate the hexagonal DNA condensate in equilibrium with a bulk solution contains 50mM NaCl, 10mM $MgCl_2$ and varying $MgSO_4$. Fugacities of each ion species used in our simulation can be found somewhere else [22]. In this way, one can study the influence of divalent coions on DNA condensation.

For each simulation run, about 500 million MC moves are carried out depending on the average number of ions in the system. To ensure thermalization, about 50 million initial moves are discarded before doing statistical analysis of the result of the simulation. All simulations are done using the physics simulation library SimEngine develop by one of the author (TTN). This library use OpenCL and OpenMP extensions of the C programming language to distribute computational workloads on multi-core CPU and GPGPU to speed up the simulation time. Both molecular dynamics and Monte-Carlo simulation methods are supported. In this paper the Monte-Carlo module of the library is used.

3. Result and Discussion

In Fig. 2 the osmotic pressure of DNA bundle at different counterion concentrations, c_{2+} is plotted as a function of the inter-axial DNA distance. Because this osmotic pressure is directly related to the "effective" force between DNA molecules at that inter-axial distance [19, 20], this figure also serves as a plot of DNA−DNA interaction. As one can see, there is a short−range attraction between two DNA molecules as they approach each other. The minimum osmotic pressure reaches as low as −5atm at high divalent counterion concentration. This is the well-known phenomenon of like-charge attraction between macroions [13, 14, 23] mediated by multivalent counterions (divalent counterion in this study). The maximum attraction occurs at the distance $d \simeq 26 - 27$Å, in good agreement with various theoretical and experimental results [4, 24]. For smaller d, the DNA-DNA interaction experiences sharp increase due to the hardcore repulsion between the counterions.

From the P-V curve, we can also can calculate the free energy, $\mu_{\rm DNA}$, of packaging DNA into bundles by integrating the pressure with the volume of the bundle. Per DNA nucleotide

IOP Conf. Series: Journal of Physics: Conf. Series **865** (2017) 012010 doi:10.1088/1742-6596/865/1/012010

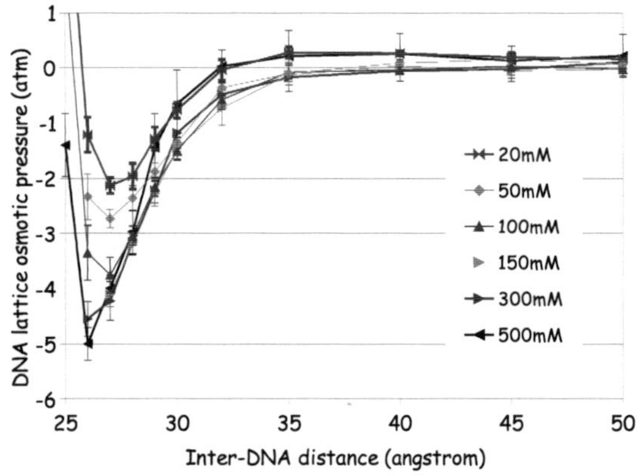

Figure 2. The osmotic pressure of the DNA bundle as function of the inter-axial DNA distance d for different divalent counterion concentration c_Z shown in the inset. The solid lines are guides to the eye. The counterion radius is 2.0 Å

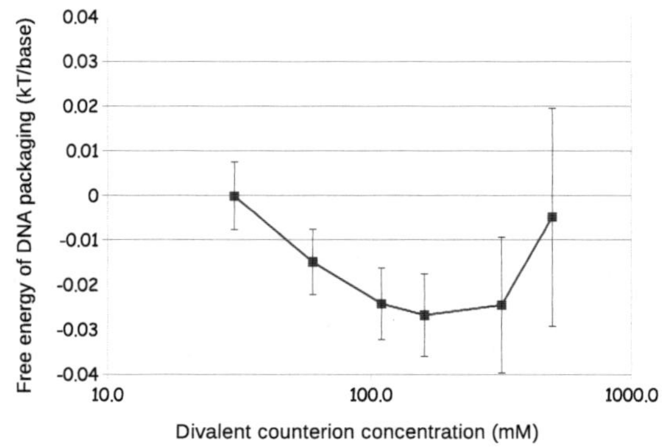

Figure 3. The free energy of packaging DNA molecule into hexagonal bundles as a function of the divalent counterion concentrations (Mg^{2+}) in the presence of divalent coions (SO_4^{2-}). The points are results of numerical integration of P_{osm} from Fig. 3.

base, the packaging free energy is given by:

$$\mu_{\rm DNA}(d) = \frac{l}{L_z N_{\rm DNA}} \int_\infty^d P_{osm}(d')dV = \frac{l}{N_{\rm DNA}} \int_\infty^d P_{osm}(d') \frac{2L_x L_y}{d'} dd' \qquad (2)$$

here $l = 1.7$Å is the distance between DNA nucleotides along the axis of the DNA. The numerical result for $\mu_{\rm DNA}(d^*)$ at the optimal bundle lattice constant d^* is plotted in Fig. 3 as function of the divalent counterion concentration c_{2+}. Fig. 3 shows a non-monotonic dependence of the electrostatic contribution to DNA packaging free energy. There is an optimal concentration, c^*, where the free energy cost of packaging DNA is lowest. It is negative indicating the tendency of the divalent counterions to condense the DNA. At smaller or larger concentrations of the counterions, the free energy cost of DNA packaging is higher. These results are consistent with the correlation theory of DNA reentrant condensation by multivalent counterions [25, 23, 10] and the experiment results on ejecting DNA from bacteriophage under varying counterion concentrations [9].

In Fig. 3, the error bar of free energy is large at high counterion concentration. This is due to the Grand Canonical Monte Carlo simulation method that we use in our work. There is a limitation of application of this method at high particle density. In this limit, the Metropolis acceptance probability of adding or removing a salt molecule is very small. This leads to the requirement of long simulation and large sampling time to obtain good statistical result. Nevertheless, the non-monotonic dependence of free energy of DNA packaging on Mg^{2+} counterion concentration has been demonstrated in experiment [9] and matches qualitatively with the average values of free energy in our simulation. We believe that the large error bar due to finite simulation time do not change this qualitative result.

Fig. 3 gives the short−range attraction among DNA molecules to be $-0.025k_BT$/base. This is larger than the fitted value obtained from the viral DNA ejection experiments [26]. There are many factors that lead to this quantitative discrepancy such as DNA flexibility. ion sizes, DNA orientations (twisting, frustrations),... [27, 28, 7]All these factors are expected to reduce the attraction between the DNA compared to our idealized simulation. Nevertheless, the non-monotonic electrostatic influence of divalent counterions on DNA-DNA "effective" interaction is clearly demonstrated in our idealized simulation. More importantly, compare to the case of no divalent coions [18], our result for NaCl + $MgCl_2$+ $MgSO_4$ mixture shows that divalent coions reduce of depth of DNA−DNA interaction by about 25% and make DNA decondensation happen sooner. This match qualitatively with experiment result of $MgCl_2$ and $MgSO_4$ solution [8, 9].

4. Conclusion

In this paper, we presented a study of the DNA condensation problem using the Grand canonical Monte−Carlo simulation using a primitive model for the mobile ions. is suppressed in simulation by fixing the position of the DNA cylinders in the bundle. Such study can be applied directly to the experimental problem of DNA ejection from bacteriophages where DNA condensed in a strongly confined environment. It is shown that divalent counterions strongly influence DNA interaction and packaging. The simulation results for divalent counterions with 2.0Å radius show that the electrostatic free energy of packaging DNA into hexagonal bundle varies non-monotonically with the counterion concentration. We study DNA condensation with different divalent co-ion concentration. It is show that divalent co-ions reduce DNA condensation free energy and make DNA de-codensation happens earlier. This can explain experimentally observed quantitive differences of DNA condensation inside viruses in the presences of $MgCl_2$ salt mixture and in the presence of $MgSO_4$ salt mixture.

5. Acknowledgments

We would like to thank Drs. A. Lyubartsev, B. Shklovskii, A. Evilevich, L. T. Fang, W. Gelbart, T. X. Hoang, Paolo Carloni for valuable discussions. TTN acknowledges the financial support of the Vietnam National Foundation for Science and Technology NAFOSTED Contract 103.02-2012.75 and the USA National Science Foundation grant NSF CBET-1134398. The authors are indebted to Dr A. Lyubartsev for providing us with the Fortran source code of their Expanded Ensemble Method.

References

[1] Surana S, Shenoy A R and Krishnan Y 2015 *Nature Nanotechnology* **10** 741–747
[2] Jones M R, Seeman N C and Mirkin C A 2015 *Science* **347** 1260901
[3] Knobler C M and Gelbart W M 2009 *Annu. Rev. Phys. Chem.* **60** 367–383
[4] Rau D C and Parsegian V A 1992 *Biophys. J.* **61** 246
[5] Hud N V and Downing K H 2001 *Proc. Nat. Acad. Sci. USA* **98** 14925–14930
[6] Hoang T X, Giacometti A, Podgornik R, Nguyen N T T, Banavar J R and Maritan A 2014 *J. Chem. Phys.* **140** 064902
[7] Grason G M 2010 *Phys. Rev. Lett.* **105** 045502
[8] Evilevitch A, Lavelle L, Knobler C M, Raspaud E and Gelbart W M 2003 *Proc. Nat. Acad. Sci. USA* **100** 9292
[9] Evilevitch A, Fang L T, Yoffe A M, Castelnovo M, Rau D C, Parsegian V A, Gelbart W M and Knobler C M 2008 *Biophys. J.* **94** 1110
[10] Nguyen T T, Rouzina I and Shklovskii B I 2000 *J. Chem. Phys.* **112** 2562
[11] Saminathan M, Antony T, Shirahata A, Sigal L H, Thomas T and Thomas T J 1999 *Biochemistry* **38** 38213830
[12] Pelta J, Durand D, Doucet J and Livolant F 1996 *Biophys. J.* **71** 48–63
[13] Naji A, Arnold A, Holm C and Netz R R 2004 *Eur. Phys. Lett.* **67** 130–136
[14] Gelbart W M, Bruinsma R F, Pincus P A and Parsegian A V 2000 *Phys. Today* **53** 38–44
[15] Koltover I, Wagner K and Safinya C R 2000 *Proc. Nat. Acad. Sci. USA* **97** 14046–14051
[16] Lee S, Le T T and Nguyen T T 2010 *Phys. Rev. Lett.* **105** 248101
[17] Nguyen T T 2013 *J. Biol. Phys.* **39** 247
[18] Nguyen T T 2016 *J. Chem. Phys.* **144** 065102

[19] Lyubartsev A P and Nordenskiöld L 1995 *J. Phys. Chem.* **99** 10373–10382
[20] Guldbrand L, Nilsson L G and Nordenskiöld L 1986 *J. Chem. Phys.* **85** 6686–6698
[21] Valleau J P and Cohen L K 1980 *J. Chem. Phys.* **72** 5935–5941
[22] Duc N V and Nguyen T T 2016 *Phys. Rev. Lett. submitted* (*Preprint* `arxiv:1608.01912v2`)
[23] Grosberg A Y, Nguyen T T and Shklovskii B 2002 *Rev. Mod. Phys.* **74** 329
[24] Purohit P K, Inamdar M M, Grayson P D, Squires T M, Kondev J and Phillips R 2005 *Biophys. J.* **88** 851–866
[25] Shklovskii B I 1999 *Phys. Rev. E* **60** 5802–5811
[26] Lee S, Tran C V and Nguyen T T 2011 *J. Chem Phys.* **134** 125104 (*Preprint* `cond-mat/0811.1296`)
[27] Grønbech-Jensen N, Mashl R J, Bruinsma R F and Gelbart W M 1997 *Phys. Rev. Lett.* **78** 24772480
[28] Lyubartsev A P, Tang J X, Janmey P A and Nordenskiöld L 1998 *Phys. Rev. Lett.* **81** 5465–5468

IOP Conf. Series: Journal of Physics: Conf. Series **865** (2017) 012011 doi:10.1088/1742-6596/865/1/012011

Microscopic description of average level spacing in even-even nuclei

Le Thi Quynh Huong[1,2,3], **Nguyen Quang Hung**[1], **and Le Tan Phuc**[1,3]

[1] Institute of Research and Development, Duy Tan University, K7/25 Quang Trung, Danang city, Vietnam

[2] Department of Natural Science and Technology , University of Khanh Hoa, Nha Trang City, Khanh Hoa Province, Vietnam

[3] Faculty of Physics and Engineering Physics, Ho Chi Minh University of Science, Ho Chi Minh City, Vietnam

E-mail: `lethiquynhhuong@ukh.edu.vn (L.T.Q.H)`

Abstract. A microscopic theoretical approach to the average level spacing at the neutron binding energy in even-even nuclei is proposed. The approach is derived based on the Bardeen-Cooper-Schrieffer (BCS) theory at finite temperature and projection M of the total angular momentum J, which is often used to describe the superfluid properties of hot rotating nuclei. The exact relation of the J-dependent total level density to the M-dependent state densities, based on which the average level spacing is calculated, was employed. The numerical calculations carried out for several even-even nuclei have shown that in order to reproduce the experimental average level spacing, the M-dependent pairing gaps as well as the exact relation of the J-dependent total level density formula should be simultaneously used.

1. Introduction

Knowledge of the average level spacing between resonances of a compound nucleus has an important role in the calculation of the nuclear level density (NLD). The latter, which is defined as the number of excited levels per unit of excitation energy E^*, has major contribution in the study of nuclear structure and low-energy reaction. This average level spacing has been extensively studied within a number of theoretical models including both phenomenological and microscopic ones (see e.g., Refs. [1, 2, 3] and references therein). The phenomenological models are often derived based on the Bethe formula of NLD [4], in which several phenomenological parameters are introduced and their values are found from the fitting to the experimental data. The microscopic models are usually constructed based on the statistical nuclear thermodynamic theories such as the Bardeen-Cooper-Schrieffer (BCS) theory at finite temperature (FTBCS) [5] or finite temperature Hartree-Fock-Bogoliubov (FTHFB) [6], in which the effects of superfulid pairing, deformation, vibration, rotation and shell structure, which are important for the description of NLD, are microscopically or empirically included. However, within the FTBCS and FTHFB, the angular-momentum dependent total NLD is calculated approximately from the state density by assuming that the distribution of nuclear spins can be approximated with the Gaussian function [7]. In fact, the angular-momentum dependent total NLD can be calculated directly from the angular-momentum dependent state densities via its exact relation (see e.g., Eq. (56) of Ref. [7]), which can be obtained within the BCS theory at finite temperature and

IOP Conf. Series: Journal of Physics: Conf. Series **865** (2017) 012011 doi:10.1088/1742-6596/865/1/012011

angular momentum (FTABCS). The goal of the present paper is to employ the FTABCS theory to calculate the angular-momentum dependent total NLD using this exact relation. The results obtained will be used to calculate the average level spacing at the neutron binding energy. The numerical calculations are restricted to even-even nuclei, whereas the extension to odd-even and odd-odd cases is still going on.

2. Formalism

2.1. BCS theory at finite temperature and finite angular momentum (FTABCS)

The FTABCS equations are derived based on the grand-canonical ensemble average of the nuclear Hamiltonian, which consists of the one-body mean field and two-body monopole pairing interaction with the rotational field acted on the projection M of the total angular momentum J as an external field. The derivation of the FTABCS equations for the pairing gap Δ, particle number N, and angular momentum M has been reported in a number of papers [7, 8, 9, 10], so we report here only their final forms, which are given as

$$\begin{aligned} \Delta &= G\sum_k u_k v_k (1 - n_k^+ - n_k^-) , \\ N &= 2\sum_k \left[(1 - n_k^+ - n_k^-) v_k^2 + \frac{1}{2}(n_k^+ + n_k^-) \right] , \\ M &= \sum_k m_k (n_k^+ - n_k^-) , \end{aligned} \tag{1}$$

where

$$\begin{aligned} u_k^2 &= \frac{1}{2}\left(1 + \frac{\epsilon_k - \lambda}{E_k}\right), \quad v_k^2 = 1 - u_k^2 , \\ E_k &= \sqrt{(\epsilon_k - \lambda)^2 + \Delta_k^2} , \quad n_k^{\pm} = \frac{1}{1 + e^{(E_k \mp \gamma m_k)/T}} . \end{aligned}$$

Here ϵ_k are energies of the single-particle levels k written in the deformed basis and λ, γ, G, and T are respectively chemical potential, rotational velocity (frequency), pairing interaction parameter, and temperature of the nuclear system, which corresponds to a given excitation energy of the compound nucleus.

The FTABCS equations are numerically solved to obtain λ, γ, Bogoliubov transformation coefficients u_k and v_k, quasiparticle energy E_k, quasiparticle occupation numbers $n_k^{\pm}$, and pairing gap $\Delta_{N(Z)}$ at different values of T and M. These quantities are then used to calculate the total (internal) energy $E(T, M)$ and entropy $S(T, M)$, whose explicit forms are given as

$$E(T, M) = 2\sum_k \epsilon_k \left[(1 - n_k^+ - n_k^-) v_k^2 + \frac{1}{2}(n_k^+ + n_k^-) \right] - \frac{\Delta^2}{G} , \tag{2}$$

$$S(T, M) = -\sum_k [n_k^+ \ln n_k^+ + (1 - n_k^+)\ln(1 - n_k^+) + n_k^- \ln n_k^- + (1 - n_k^-)\ln(1 - n_k^-)] . \tag{3}$$

As the nuclear system consists of N neutrons and Z protons, the total energy and total entropy of the nucleus should be calculated as

$$E(T, M) = E_N(T, M) + E_Z(T, M), \quad S(T, M) = S_N(T, M) + S_Z(T, M), \tag{4}$$

where $E_{N(Z)}$ and $S_{N(Z)}$ are the internal energy and entropy calculated from Eqs. (2) and (3) for neutron (proton), respectively.

2.2. Intrinsic state density and average level spacing

The intrinsic state density of an atomic nucleus at given excitation energy E^* and angular momentum M is calculated based on the inverse Laplace transform of the grand-partition function making use of the saddle-point approximation. It has the form as [7]

$$\omega(E^*, M) = \frac{e^S}{(2\pi)^2 |D|^{1/2}} , \tag{5}$$

where $E^*(T, M) = E(T, M) - E(T = 0, M)$ and the determinant D is given in terms of the second derivatives of the grand-partition function with respect to $\alpha = \lambda/T$ and $\mu = \gamma/T$

$$D = \begin{vmatrix} \frac{\partial^2\Omega}{\partial\alpha_N^2} & \frac{\partial^2\Omega}{\partial\alpha_N\partial\alpha_Z} & \frac{\partial^2\Omega}{\partial\alpha_N\partial\alpha_\mu} & \frac{\partial^2\Omega}{\partial\alpha_N\partial\alpha_\beta} \\ \frac{\partial^2\Omega}{\partial\alpha_Z\partial\alpha_N} & \frac{\partial^2\Omega}{\partial\alpha_Z^2} & \frac{\partial^2\Omega}{\partial\alpha_Z\partial\alpha_\mu} & \frac{\partial^2\Omega}{\partial\alpha_Z\partial\alpha_\beta} \\ \frac{\partial^2\Omega}{\partial\alpha_\mu\partial\alpha_N} & \frac{\partial^2\Omega}{\partial\alpha_\mu\partial\alpha_Z} & \frac{\partial^2\Omega}{\partial\alpha_\mu^2} & \frac{\partial^2\Omega}{\partial\alpha_\mu\partial\alpha_\beta} \\ \frac{\partial^2\Omega}{\partial\alpha_\beta\partial\alpha_N} & \frac{\partial^2\Omega}{\partial\alpha_\beta\partial\alpha_Z} & \frac{\partial^2\Omega}{\partial\alpha_\beta\partial\alpha_\mu} & \frac{\partial^2\Omega}{\partial\alpha_\beta^2} \end{vmatrix} . \tag{6}$$

In Eq. (6), the logarithm of the grand-partition function of the systems Ω is calculated as

$$\Omega = \Omega_N + \Omega_Z = S_N + S_Z + \alpha N_N + \alpha Z_Z + \mu M - \beta E . \tag{7}$$

The derivations of Ω with respect to α and μ can be seen explicitly in Eqs. (25)–(35) of Ref. [7].

The total NLD $\rho(E^*)$ at given excitation energy E^* is calculated as the sum of all J-dependent level densities

$$\rho(E^*) = \sum_J (2J + 1)\rho(E^*, J) , \tag{8}$$

where $\rho(E^*, J)$ can be obtained from the difference between the state densities of the system with angular momentum $M = J$ and $M = J + 1$, namely [7]

$$\rho(E^*, J) = \omega(E^*, M = J) - \omega(E^*, M = J + 1) , \tag{9}$$

with $\omega(E^*, M)$ being the state density calculated from Eq. (5). It is worth noticing here that Eq. (9) is the exact relation between the total J-dependent NLD and M-dependent state densities.

Based on the assumption that the distribution of nuclear spin can be approximately expressed in terms of the Gaussian function, $\rho(E^*, J)$ in Eq. (8) can be calculated approximately as [5, 7]

$$\rho(E^*, J) \approx \frac{2J + 1}{2\sigma^3\sqrt{2\pi}}\omega(E^*)\exp\left(-\frac{J(J + 1)}{2\sigma^2}\right) , \tag{10}$$

where $\omega(E^*)$ is the total state density obtained from Eq. (5) with $M = 0$ and σ is the spin cut-off parameter, which is related to the nuclear moment of inertia via the relation $\sigma^2_{\perp(\parallel)} = \Im_{\perp(\parallel)}T/\hbar$, where $\Im_{\perp(\parallel)}$ is the moment of inertia perpendicular (parallel) to the symmetry axis of the nucleus. The perpendicular spin cut-off parameter is often calculated empirically via the limit of rigid body [11]

$$\sigma^2_\perp \approx 0.015TA^{5/3}(1 + \frac{\beta_2}{3}) , \tag{11}$$

where A is the nuclear mass number and β_2 is the quadrupole deformation parameter. The parallel spin cut-off parameter can be calculated microscopically within the FTBCS theory as [7]

$$\sigma^2_\parallel = \frac{1}{2}\sum_k m_k^2 \text{sech}^2\frac{1}{2}\beta E_k . \tag{12}$$

IOP Conf. Series: Journal of Physics: Conf. Series **865** (2017) 012011 doi:10.1088/1742-6596/865/1/012011

Consequently, two approximate formulas for the J-dependent total NLD, which have been used in most theoretical models such as the FTBCS or FTHFB, are [5, 6]

$$\rho_{\rm vib}(E^*, J) \approx \frac{2J+1}{2\sigma_{\parallel}^3\sqrt{2\pi}}\frac{\omega(E^*)}{(1-e^{-\beta\omega_\lambda})^{2\lambda+1}}\exp\left(-\frac{J(J+1)}{2\sigma_{\parallel}^2}\right), \tag{13}$$

$$\rho_{\rm rot}(E^*, J) \approx \frac{2}{2\sigma_{\parallel}\sqrt{2\pi}}\omega(E^*)\sum_{K=-J}^{J}\exp\left(-\frac{K^2}{2\sigma_{\parallel}^2}-\frac{J(J+1)-K^2}{2\sigma_{\perp}^2}\right), \tag{14}$$

where ω_λ is the phonon energy of the vibrational excitation corresponding to the phonon multipolarity λ, whereas K is the projection of total angular momentum J on the symmetry axis. In these equations, Eq. (13) is used for spherical and/or slightly deformed nuclei, whose vibrational collective vibrations are significant, whereas Eq. (14) is applicable for well-deformed nuclei, whose rotational excitations are mostly important.

Knowing the J-dependent total NLD, one can easily calculate the average level spacing $\overline{D}$ at the neutron binding energy B_n based on its definition as [6]

$$\overline{D} = \frac{10^6}{\rho\left(B_n, I_t-\frac{1}{2}\right)+\rho\left(B_n, I_t+\frac{1}{2}\right)}, \tag{15}$$

where $\rho(B_n, I_t \pm 1/2)$ are the NLDs obtained at the excitation energy $E^* = B_n$ and spins $J = I_t \pm 1/2$, with I_t being the ground-state spin of the target nucleus $(Z, N-1)$.

3. Numerical results and discussion

The average level spacings are calculated within the FTABCS for several even-even nuclei from the medium ^{58}Fe to the heavy ^{250}Cf mass isotopes. The single-particle spectra ϵ_k are taken from the axially deformed Woods-Saxon (WS) potential including the spin-orbit and Coulomb interactions. The parameters of the WS potential are taken from Ref. [12]. The quadrupole β_2 and hexadecapole β_4 deformation parameters are adjusted to reproduce the experimental ground-state properties including nuclear binding energy and charge radii as well as energy of the first 2^+ state. The pairing interaction parameter $G_{N(Z)}$ is as usual adjusted so that the pairing gap $\Delta_{N(Z)}$ obtained within the FTABCS at $T = 0$ and $M = 0$ fits the experimental odd-even mass difference.

It is found that the pairing gap $\Delta_{N(Z)}$ decreases with increasing both T and M and collapses at given $T = T_c$ and $M = M_c$ values, which are defined as the critical temperature and critical angular momentum, respectively (See e.g., Figs. 2 and 3 of Ref. [8]). This change of the pairing gaps is known to affect the state density (5) as well as level density (9) at low E^* and M. Consequently, one can see in Table 1 the average level spacings $\overline{D}$ obtained within the FTABCS using Eq. (15) with $\rho(E^*, J)$ being calculated from its exact relation (9) are in quite good agreement with the experimental data of all nuclei under consideration. It is also seen in this Table 1 that the results obtained within the FTBCS by Maino [5] and FTHFB by Goriely [6] are far from the measured data, especially for nuclei with large ground-state spins such as ^{68}Zn($I_t = 5/2$), ^{88}Sr($I_t = 9/2$), ^{98}Mo($I_t = 5/2$), ^{144}Nd($I_t = 7/2$), ^{150}Sm($I_t = 7/2$), ^{162}Dy($I_t = 5/2$), ^{168}Er($I_t = 7/2$), ^{248}Cm($I_t = 9/2$), and ^{250}Cf($I_t = 9/2$). One of the reasons of the discrepancy between the results obtained within the FTBCS and FTHFB approaches and the experimental data is that the pairing gaps obtained within these theoretical models depend only on $T(E^*)$, whereas in reality the pairing gaps should depend on not only $T(E^*)$ but also $M(J)$. Other reason is certainly due to the use of the approximate formulas (13) and (14) within the FTBCS (FTHFB) instead of the exact relation (9) for $\rho(E^*, J)$.

IOP Conf. Series: Journal of Physics: Conf. Series **865** (2017) 012011 doi:10.1088/1742-6596/865/1/012011

Compound Nucleus	I_t	B_n	$\overline{D}_{exp}$	$\overline{D}_{Maino}$	$\overline{D}_{Goriely}$	$\overline{D}_{FTABCS}$
^{58}Fe	$\frac{1}{2}$	10.044	7050 ± 700		3440	6889
^{68}Zn	$\frac{5}{2}$	10.198	370 ± 20		235	365.7
^{78}Se	$\frac{1}{2}$	10.498	120 ± 15		409	122.6
^{88}Sr	$\frac{9}{2}$	11.112	290 ± 80		160	285.1
^{98}Mo	$\frac{5}{2}$	8.643	60 ± 10	30.3	42	59.96
^{112}Cd	$\frac{1}{2}$	9.394	27 ± 2	23.9	14.1	26.86
^{118}Sn	$\frac{1}{2}$	9.324	61 ± 7	76.7	38.1	68.58
^{126}Te	$\frac{1}{2}$	9.113	43 ± 3	49.0	20.8	47.14
^{132}Xe	$\frac{3}{2}$	8.935	49 ± 8	52.2	34.1	48.11
^{144}Nd	$\frac{7}{2}$	7.817	38 ± 2	30.1	25.4	34.49
^{150}Sm	$\frac{7}{2}$	7.987	2.4 ± 0.2	1.56	1.94	2.486
^{162}Dy	$\frac{5}{2}$	8.197	2.4 ± 0.2	2.67	1.48	2.464
^{168}Er	$\frac{7}{2}$	7.771	4.2 ± 0.3	4.61	2.7	3.987
^{180}Hf	$\frac{9}{2}$	7.380	4.6 ± 0.3	3.73	3.56	6.768
^{202}Hg	$\frac{3}{2}$	7.754	90 ± 30		36.4	86.75
^{248}Cm	$\frac{9}{2}$	6.21	1.27 ± 0.19	1.48	1.28	1.242
^{250}Cf	$\frac{9}{2}$	6.62	0.79 ± 0.12	0.73	0.426	0.7738

Table 1. Average level spacings of even-even nuclei at the neutron binding energy (B_n) and the ground-state spin (I_t) of the target nucleus obtained within the FTABCS ($\overline{D}_{FTABCS}$) in comparison with the experimental data ($\overline{D}_{exp}$) [14] as well as the FTBCS given by Maino ($\overline{D}_{Maino}$) [5] and FTHFB given by Goriely ($\overline{D}_{Goriely}$) [6].

4. Conclusions

Present paper applies the BCS theory at finite temperature and angular momentum (FTABCS) to describe the average level spacings of several even-even nuclei. Within the FTABCS, the neutron and proton pairing gaps and total state densities, which depend on both temperature (excitation energy) and angular momentum, are microscopically calculated. The average level spacings are then obtained based on the exact relation of the angular-momentum dependent total level density to the state densities instead of the approximate formulas, which are often used in previously proposed theoretical approaches. The results obtained within the FTABCS are in good agreement with the experimental data, especially for nuclei having large ground-state spins, whose angular-momentum dependent pairing correlations are important. Present paper is restricted to even-even nuclei. The extension of this method will be studied and the results will be reported in the forthcoming papers.

IOP Conf. Series: Journal of Physics: Conf. Series **865** (2017) 012011 doi:10.1088/1742-6596/865/1/012011

References

[1] Lynn J E 1968 *The theory of neutron resonance reactions*(Clarendon Press, Oxford); Baba H 1970 *Nucl. Phys.* **A 159** 625; Dilg W *et. al.* 1973 *Nucl. Phys.* **A217** 269.
[2] Huizenga J R *et. al.* 1974 *Nucl. Phys.* **A223** 577.
[3] Huizenga J R and Moretto L G 1972 *Ann. Rev. Nucl. Sci.* **22** 427.
[4] Newton T P 1956 *Can. J. Phys.* **34** 804; Cameron A G W 1958 *Can. J. Phys.* **36** 1040; Gilbert A and Cameron A G W 1965 *Can. J. Phys.* **43** 1446; Ignatyuk A V, Smirenkin G N, and Tishin A S 1975 *Sov. J. Nucl. Phys.* **21** 255.
[5] Maino G, Menapace E, and Ventura A 1980 *Nuovo Cimento* **A57** 427; Benzi V, Maino G, and Menapace E 1981 *Nuovo Cimento* **A66** 1.
[6] Goriely S 1996 *Nucl. Phys.* **A605** 28; Goriely S, Hilaire S, and Koning A J 2008 *Phys. Rev.* **C78** 064307; Hilaire S and Goriely S 2006 *Nucl. Phys.* **A779** 63.
[7] Moretto L G 1972 *Nucl. Phys.* **A185** 145.
[8] Hung N Q and Dang N D 2008 *Phys. Rev.* **C78** 064315.
[9] Hung N Q and Dang N D 2011 *Phys. Rev.* **C84** 054324.
[10] Huong L T Q, Hung N Q and Trang L T Q 2016 *Jour. Phys.: Conf. Ser.* **726** 012011.
[11] Dilg W, Schantl W, Vonach H, and Uhl M 1973 *Nucl. Phys.* **A217** 269.
[12] Cwiok S *et al.* 1987 *Comput. Phys. Commun.* **46** 379.
[13] Bohr A and Mottelson B 1996, *Nuclear Structure*, Vol. 2, Benjamin, New York.
[14] https://www-nds.iaea.org/RIPL-3/.

IOP Conf. Series: Journal of Physics: Conf. Series **865** (2017) 012012 doi:10.1088/1742-6596/865/1/012012

A non–extensive thermodynamic theory of ecological systems

Le Van Xuan[1], Nguyen Khac Ngoc[2], Nguyen Tri Lan[3] and Nguyen Ai Viet[3]

[1]Hanoi College of Technology and Economics, Soc Son, Ha Noi, Viet Nam

[2]Hung Yen Industrial College, Hung Yen Province, Viet Nam

[3]Institute of Physics,Vietnam Academy of Science and Technology, 10 Dao Tan, Ha Noi, Viet Nam

E-mail: xuanlv.hn@gmail.com

Abstract. After almost 30 years of development, it is not controversial issue that the so–called Tsallis entropy provides a useful approach to studying the complexity where the non–additivity of the systems under consideration is frequently met. Also, in the ecological research, Tsallis entropy, or in other words, q–entropy has been found itself as a generalized approach to define a range of diversity indices including Shannon–Wiener and Simpson indices. As a further stage of development in theoretical research, a thermodynamic theory based on Tsallis entropy or diversity indices in ecology has to be constructed for ecological systems to provide knowledge of ecological macroscopic behaviors. The standard method of theoretical physics is used in the manipulation and the equivalence between phenomenological thermodynamics and ecological aspects is the purpose of the ongoing research.

The present work is in the line of the authors research to implement Tsallis non–extensivity approach to obtain the most important thermodynamic quantities of ecological systems such as internal energy U_q and temperature T_q based on a given modeled truncated Boltzmann distribution of the Whittaker plot for a dataset. These quantities have their own ecological meaning, especially the temperature T_q provides the insight of equilibrium condition among ecological systems as it is well–known in 0th law of thermodynamics.

1. Introduction

In the history of science, the applications of thermodynamics and statistical physics, and their methods on research of complex systems, especially biological and ecological systems is not a new trend in interdisciplinary science [1–3]. The basic of these branches of physics brings new methodologies, new ideas and new understanding into ecological science [3–7]. In its birth, thermodynamics was considered as a phenomenological and experimental science trying to describe the macroscopic states of many–body physical systems with minimal sets of observable macroscopic variables. And only since the beginning of the last century, the nature of thermodynamical processes and properties were understood and explained from the points of view of kinetic theory and statistics which were pioneered by the works of Gibbs and Boltzmann. That statistical background requires to determine and to define a set of identical microscopic particles involving in physical processes within systems. These different basic principles have caused a lot of discussions in applicability of thermodynamics and statistical physics to complex

IOP Conf. Series: Journal of Physics: Conf. Series **865** (2017) 012012 doi:10.1088/1742-6596/865/1/012012

systems such as biological, ecological, social and others, where the definitions of macroscopic and microscopic objects are depending not only on systems theirselves but also on the scales of observation. In despite of these differences in understanding the attempts of searching new applications of thermodynamics and statistical physics in the research of complex systems are always attracting the attention of many researchers in many aspects, from theory to practice.

To continue these great ideas, the present work as an extension of the previous work [8] focuses on finding the way of constructing a thermodynamic theory for ecological systems. However, instead of attempting to define the interacting individuals in a ecological system, the approach used in the present work tries to build up a thermodynamic consistent theory from the diversity index which is the most common ecological quantity obtained from ecological measurements, and within the proposed approach the most important thermodynamic quantities such as internal energy U and temperature T are explicitly defined and obtained. It would be expected that the results should cover in itself ecological meaning and also fulfill the thermodynamic consistence of physics.

2. Tsallis entropic approach to a thermodynamic theory

Thermodynamic is full of highly scientific and charming terms and concepts, giving an impression of philosophical and scientific profundity. Entropy, thermal death of the Universe, ergodicity, statistical ensemble—all these words sound very impressive posed in any order. But, placed in the appropriate order, they can help us to find the solution of urgent practical problems.

Unlike the most of traditional physical systems where the non–extensivity is very small, even gets zero in every experimental observations, ecological systems or in general, complex systems have visible non—extensivity, in same cases, very strong [8–12]. Furthermore, experimental observations in ecological systems are indirect in whole, and always in situations of information incompleteness. Due to these difficulties, a new physical and mathematical tool has to be developed to take into account of the non–extensivity, and the Tsallis entropy [13] would be the best nominee available in this scope of works [8].

2.1. Non–extensive statistical mechanics

A non–extensive statistical mechanics of a complex system will be built upon the Tsallis entropy [9, 13–16]

$$S_q = \frac{1}{q-1}\left(1-\sum_{i=1}^{W} p_i^q\right), \tag{1}$$

where $q \in R$ is some real parameter characterizing the non–extensivity, W is the number of possible microscopic states, and normalized constraint of statistical weight p_i

$$\sum_{i=1}^{W} p_i = 1. \tag{2}$$

To construct statistical mechanics from the above given entropy S_q, it has to be optimized with the normalized constraint (2) to statistical weight p_i and some constraint related to internal energy [11–13, 15]

$$\sum_{i=1}^{W} p_i E_i = U, \tag{3}$$

where $\{E_i\}$ are the eigenvalues of some Hamiltonian H with the appropriate boundary conditions, and the internal energy U is a finite fixed value. The constraint (3) would be generalized in non–extensive manner as

$$\begin{aligned}\langle \mathcal{H} \rangle_q &= \left(\sum_{i=1}^{W} p_i^q \right)^{-1} \sum_{i=1}^{W} p_i^q E_i \\ &= U_q,\end{aligned} \tag{4}$$

with $\langle \ldots \rangle_1 = \langle \ldots \rangle$ and $\left(\sum_{i=1}^{W} p_i^q \right)^{-1} p_i^q$ is referred to as the *escort distribution*. The optimization of the Tsallis entropy (1) with the constraints (2) and (4) straightforwardly yields

$$p_i = \left(\bar{Z}_q \right)^{-1} \mathrm{e}_q^{-\beta_q (E_i - U_q)}, \tag{5}$$

with

$$\bar{Z}_q = \sum_{i=1}^{W} \mathrm{e}_q^{-\beta_q (E_i - U_q)}, \tag{6}$$

and

$$\beta_q = \left(\sum_{i=1}^{W} p_i^q \right)^{-1} \beta, \tag{7}$$

where $\beta = (kT)^{-1}$ being the Lagrange parameter associated with constraint (4). Here to avoid confusion, the parameter k, which should be the Boltzmann constant k_{B} in the context of standard statistical mechanics, has been restored. And the q–exponential function e_q^x is defined as

$$\mathrm{e}_q^x = (1 + (1-q)\, x)^{\frac{1}{1-q}}, \tag{8}$$

that approaches to the exponential function when q goes to 1,

$$\lim_{q \to 1} \mathrm{e}_q^x = \mathrm{e}_q^x. \tag{9}$$

2.2. Connection to q–Thermodynamics

Following the standard routine of thermodynamics presented in original works and textbooks [13–15, 17, 18], it would be easy to define an abstract temperature T within q–thermodynamics as

$$\frac{1}{T} = \frac{\partial S_q}{\partial U_q}, \tag{10}$$

which is identical to physical temperature in normal situations when q approaches to 1. Other q–thermodynamic potentials such as free energy F_q

$$F_q = U_q - T S_q - kT \ln_q Z_q, \tag{11}$$

are also obtained in standard way of thermodynamics, where

$$\ln_q Z_q = \ln_q \bar{Z}_q - \beta U_q, \tag{12}$$

with the inverse of q–exponential function e_q^x

$$\ln_q x = \frac{x^{1-q} - 1}{1-q}. \tag{13}$$

To the end of this section, it would be said that all above consideration of non–extensive statistical mechanics and corresponding derivative q–thermodynamics provides a sufficiently plain physical and mathematical tool to develop a thermodynamic theory for ecological systems where the diversity indices is well determined in the manner of real datasets in ecological measurements.

3. An application to ecological systems

3.1. Truncated exponential distribution and diversity index

In very recent research [8], the rank abundance curve of a given ecological dataset [19, 20] has been proposed to be modeled by a truncated exponential distribution

$$\begin{aligned} p_{(x,\lambda)} &= p_i \\ &= \begin{cases} \frac{\exp\{\frac{C}{\lambda}\}}{\lambda(\exp\{\frac{C}{\lambda}\}-1)} \exp\left\{-\frac{x_i}{\lambda}\right\} & \text{for } 0 \le x_i \le C \\ 0 & \text{otherwise,} \end{cases} \end{aligned} \tag{14}$$

where x_i is the ith species in the community, C is maximum value of species rank and λ is some rate parameter of the rank abundance curve under consideration. The model parameters C and λ are directly obtained from a given observed dataset by fitting techniques. The diversity index would be identified as the Tsallis entropy with respect to modeled truncated exponential distribution above, and has been analytically obtained in [8]

$$S_q = \frac{1}{q-1}\left(1 - \frac{\lambda}{q}\left(\mathrm{e}^{\frac{Cq}{\lambda}} - 1\right)\left(\lambda\left(\mathrm{e}^{\frac{C}{\lambda}} - 1\right)^{-q}\right)\right). \tag{15}$$

It is easy to verify the fallback of the Tsallis entropy (15) to Boltzmann entropy in statistical mechanics and to Shanon–Wiener diversity index when parameter C approaches to infinity and non–extensivity equals to one [8]. This ensures the validity of the proposed approach.

3.2. q–Thermodynamics of the truncated exponential distribution

The internal energy of the system under consideration is obtained by performing directly analytical manipulations on expression (3)

$$U_q = \frac{\lambda}{q} - C\left(\mathrm{e}^{\frac{Cq}{\lambda}} - 1\right)^{-1}. \tag{16}$$

And, the inverse of temperature (10) is, therefore, calculated by standard routine of thermodynamics as

$$\begin{aligned} \beta_q(\lambda, C) &= \frac{\partial S_q(\lambda, C)}{\partial U_q(\lambda, C)} \\ &= \frac{\partial S_q}{\partial \lambda}\left(\frac{\partial U_q}{\partial \lambda}\right)^{-1} + \frac{\partial S_q}{\partial C}\left(\frac{\partial U_q}{\partial C}\right)^{-1}. \end{aligned} \tag{17}$$

IOP Conf. Series: Journal of Physics: Conf. Series **865** (2017) 012012 doi:10.1088/1742-6596/865/1/012012

The resultant expression of the inverse of temperature (17) in above calculation is quite long and verbose to write down here. However, its limit when C approaches to infinity is subject of further interests, because of its simplicity and ease to compare to physical well–known consequences where the number of particle could be assumed to be infinity. In the interested limit, $C \to \infty$, the inverse of temperature reads

$$\lim_{C\to\infty} \beta_q \quad = \lambda^{-q}. \tag{18}$$

The formula (18) provides the influence of non–extensivity on the physical temperature that leads to change the temperature as a q–power function of physical temperature. When the systems are extensive, or in other words, $q = 1$, the temperature (18) recovers real physical temperature as it is always in Gibbs–Boltzmann approach of thermodynamics, i.e. $k_{\rm B}T = \lambda$.

In fact, the importance of (18) in ecological research is to determine a equilibrium parameter among systems based on their diversity indices, that satisfies thermodynamic consistence. This parameter is first obtained and introduced in ecological research.

4. Conclusions

Along with previous work [8], the present work is obviously an attempt to bring thermodynamics and statistical mechanics into the field of ecological research.

What has been done in this work is to obtain the internal energy U_q and the abstract temperature T_q of a ecological system in the context of Tsallis entropy which has been proved to be considered as diversity index of the corresponding ecological system. Once, the internal energy U_q and the abstract temperature T_q of a system are known, its thermodynamic theory would be well defined. The first introduced abstract temperature T_q in ecological research plays the role of new indicator for the equilibrium condition in ecology, which is similar to real physical temperature in the 0th law of thermodynamics defining the equilibrium among systems. The results in this work also confirm the thermodynamic fulfillment of newly constructed thermodynamical theory that early developed theories have been hardly achieved.

And, the derivative thermodynamic quantities such as other thermodynamic potentials, heat capacity, and many others will be implemented in next works, to evaluate the theory and to examine their ecological meaning.

Acknowledgments

This research is funded by Vietnam National Foundation for Science and Technology Development (NAFOSTED) under grant number 103.01-2015.42.

References

[1] Young G L 1974 *Advances in Ecological Research* **8** 1–105
[2] Goerner S J 1994 *World Futures* **42** 193–214
[3] Jørgensen S E and Svirezhev Y M 2004 *Towards a Thermodynamic Theory for Ecological Systems* (Oxford: Pergamon)
[4] Anand M, Gonzalez A, Guichard F, Kolasa J and Parrott L 2010 *Diversity* **2** 395
[5] Davic R D and Welsh H H 2004 *Annual Review of Ecology and Systematics* **35** 405–434
[6] Jørgensen S E and Svirezhev Y M 2004 Chapter 13 - Application of exergy as ecological indicator and goal function in ecological modelling *Towards a Thermodynamic Theory for Ecological Systems* ed Jørgensen S E and Svirezhev Y M (Oxford: Pergamon) pp 325 – 349
[7] Colinvaux P A 1973 *Introduction to Ecology* (Wiley)
[8] Xuan L V, Lan N T and Viet N A 2016 *Journal of Physics: Conference Series* **726** 012024
[9] Tsallis C 2002 *Chaos, Solitons & Fractals* **13** 371 – 391 classical and Quantum Complexity and Non-Extensive Thermodynamics

[10] Harte J, Zillio T, Conlisk E and Smith A B 2008 *Ecology* **89** 2700–2711
[11] Harte J 2011 *Maximum Entropy and Ecology: A Theory of Abundance, Distribution, and Energetics* Oxford Series in Ecology and Evolution (Oxford University Press)
[12] Harte J and Newman E A 2014 *Trends in Ecology & Evolution* **29** 384 – 389
[13] Tsallis C 1988 *Journal of Statistical Physics* **52** 479–487
[14] Tsallis C, Mendes R S and Plastino A 1998 *Physica A: Statistical Mechanics and its Applications* **261** 534 – 554
[15] Tsallis C 2009 *Brazilian Journal of Physics* **39** 337–356
[16] Jizba P and Arimitsu T 2006 *Physica A: Statistical Mechanics and its Applications* **365** 76 – 84 Fundamental Problems of Modern Statistical Mechanics, Proceedings of the 3rd International Conference on "News, Expectations and Trends in Statistical Physics"
[17] Tsallis C 1995 *Chaos, Solitons & Fractals* **6** 539 – 559 Complex Systems in Computational Physics
[18] Balasis G, Donner R V, Potirakis S M, Runge J, Papadimitriou C, Daglis I A, Eftaxias K and Kurths J 2013 *Entropy* **15** 4844
[19] Whittaker R H 1965 *Science* **147** 250–260
[20] Whittaker R H, Levin S A and Root R B 1973 *The American Naturalist* **107** 321–338

IOP Conf. Series: Journal of Physics: Conf. Series **865** (2017) 012013 doi:10.1088/1742-6596/865/1/012013

On the mass enhancement of black body background fluctuations

Nguyen Van Hoa[1], Le Van Xuan[2], Nguyen Tri Lan[3], and Nguyen Ai Viet[3]

[1]Hochiminh Pedagogical University, 270 An Duong Vuong, Hochiminh City, Viet Nam

[2]Hanoi College of Technology and Economics, Soc Son, Ha Noi, Viet Nam

[3]Institute of Physics,Vietnam Academy of Science and Technology, 10 Dao Tan, Ha Noi, Viet Nam

E-mail: xuanlv.hn@gmail.com

Abstract. Motivated by the mass enhancement in the toy model of moving particle in Boltzmann and Gaussian background fluctuations, the contribution of black body background fluctuation to the effective mass of massless and massive particles is considered in this work. As the black body radiation depends only on its temperature, the dependence of the effective mass on the temperature is obtained. The results, therefore, provide several physical insights for the research of complex systems where the interaction and equilibrium of a system and its surroundings are still not clear. By interpreting a characteristic parameter of the environmental contribution as its “effective temperature”, “the thermal equilibrium” condition for complex systems would be discussed in the context of a thermodynamic theory.

1. Introduction

Considering the toy models of Boltzmann and Gaussian fluctuations has shown that observing and analyzing the behaviors of a particle or a wave moving in the field of fluctuations would provide a lot of information of statistical properties of the field. “In physics, a black body would be defined as an idealized physical body that absorbs all incident electromagnetic radiation, regardless of frequency or angle of incidence. A white body is one with a "rough surface [that] reflects all incident rays completely and uniformly in all directions. A black body in thermal equilibrium (that is, at a constant temperature) emits electromagnetic radiation called black body radiation. The radiation is emitted according to Planck’s law, meaning that it has a spectrum that is determined by the temperature alone, not by the body’s shape or composition.”[1] In the research of complex systems where the influence of surrounding environment is unknown, the fluctuation of the background would be characterized in terms of an effective temperature, that of a black body that would emit the same total flux of electromagnetic energy. As next stage of investigation, the black body radiation is considered as more real model of fluctuations.

IOP Conf. Series: Journal of Physics: Conf. Series **865** (2017) 012013 doi:10.1088/1742-6596/865/1/012013

2. Energy and Effective Mass in Planck's Fluctuations

2.1. Definitions of Energy and Effective Mass

In conventional understanding of modern physics, motion of a particle is characterized by its dispersion law $\mathcal{E}(p)$ where the dependence of $\mathcal{E}(p)$ on p reflects whether particle is massive or massless.In one hand, the massless particle has dispersion law in form

$$\mathcal{E}(p) = cp, \tag{1}$$

and in other hand, the massive particle m_0 has the dispersion law written in the relativistic form[2, 3] as

$$\mathcal{E}(p) = \sqrt{(mc^2)^2 + (cp)^2}, \tag{2}$$

which tends to additive law of energy

$$\mathcal{E}(p) = mc^2 + \frac{1}{2m}p^2 + \underbrace{V(p)}_{o(p^4)}, \tag{3}$$

when momentum p approaches to zero, where c is some maximal velocity of the massless motion in given media. It is important to think of a possibility that the coordinate representation of function $V(p)$

$$V(r) = \frac{1}{\sqrt{V_\Omega}} \int_\Omega \mathrm{d}p\ \mathrm{e}^{ipr} V(p), \tag{4}$$

would contain some physical insights of potential affecting on the motion of a massive particle in the medium, or in other words, interaction between massive particle and surrounding environment.

Following the standard mathematical routine found in textbooks, the mass of a moving particle with given dispersion law $\mathcal{E}(p)$ is determined by second order derivative of $\mathcal{E}(p)$ with respect to p [4, 5] as

$$\frac{1}{m^*} = \left.\frac{\mathrm{d}^2}{\mathrm{d}p^2}\mathcal{E}(p)\right|_{p=0}. \tag{5}$$

When a particle moves under influence of some fluctuations which changes the dispersion law of the particle, energy of massive particle reads[6, 7, 8]

$$\mathcal{E}(p, \epsilon) = \sqrt{(mc^2)^2 + c^2p^2 + \epsilon^2}, \tag{6}$$

then the energy of moving particle is changed, and obtained by including all contributions of fluctuations via integrating out with fluctuation distribution function $\mathcal{D}(\epsilon)$

$$\mathcal{E}[cp; \mathcal{D}] = \int_\Omega \mathrm{d}\epsilon \mathcal{D}(\epsilon) \sqrt{(mc^2)^2 + c^2p^2 + \epsilon^2}. \tag{7}$$

The effective mass of moving particle will be determined by the definition (5)

2.2. The Energy Density Function of Black Body Fluctuations

The law of back body radiation developed by Max Planck in 1900 has shown that, expressed as an energy distribution, it is the unique stable distribution for radiation in thermodynamic equilibrium. This law can be rewritten in the term of energy density function as

$$\begin{aligned} \mathcal{D}(\epsilon, k_{\mathrm{B}}T) &= \frac{1}{\mathcal{C}} \frac{2\epsilon^3}{(2\pi\hbar)^3 c^2} \frac{1}{e^{\frac{\epsilon}{k_{\mathrm{B}}T}} - 1} \\ &= \frac{15}{\pi^4 (k_{\mathrm{B}}T)^4} \frac{\epsilon^3}{e^{\frac{\epsilon}{k_{\mathrm{B}}T}} - 1} \end{aligned}, \tag{8}$$

$$\begin{aligned} \mathcal{C} &= \int_0^\infty \mathrm{d}\epsilon \frac{2\epsilon^3}{(2\pi\hbar)^3 c^2} \frac{1}{e^{\frac{\epsilon}{k_{\mathrm{B}}T}} - 1} \\ &= \frac{2}{(2\pi\hbar)^3 c^2} \int_0^\infty \mathrm{d}\epsilon \epsilon^3 \frac{1}{e^{\frac{\epsilon}{k_{\mathrm{B}}T}} - 1} \\ &= \frac{2}{(2\pi\hbar)^3 c^2} (k_{\mathrm{B}}T)^4 \int_0^\infty \mathrm{d}\left(\frac{\epsilon}{\mathrm{k_B T}}\right) \left(\frac{\epsilon}{k_{\mathrm{B}}T}\right)^3 \frac{1}{e^{\frac{\epsilon}{k_{\mathrm{B}}T}} - 1} \\ &= \frac{2}{(2\pi\hbar)^3 c^2} (k_{\mathrm{B}}T)^4 \int_0^\infty \mathrm{d}x \frac{x^3}{e^x - 1} \\ &= \frac{2}{(2\pi\hbar)^3 c^2} (k_{\mathrm{B}}T)^4 \frac{\pi^4}{15} \\ &= \frac{1}{15} \frac{2\pi}{(2\hbar)^3 c^2} (k_{\mathrm{B}}T)^4 \end{aligned} \tag{9}$$

where $\hbar\omega$ of electromagnetic radiation is replaced by the background fluctuation ϵ, and T is some effective temperature which would be generalized by an abstract temperature characterizing the equilibrium of background, in case of complex systems

2.3. Energy and Effective Mass

Performing the integrals (6) and (7) with energy density function (8) leads to the total energy

$$\begin{aligned} \varepsilon(cp, k_B T) &= \frac{15}{\pi^4 (k_{\mathrm{B}}T)^4} \int_0^\infty \mathrm{d}\epsilon \frac{\epsilon^3}{e^{\frac{\epsilon}{k_{\mathrm{B}}T}} - 1} \sqrt{(m_0 c^2)^2 + c^2 p^2 + \epsilon^2} \\ &= \sqrt{(m_0 c^2)^2 + c^2 p^2} \mathcal{I}_{1/2} \left(\frac{k_{\mathrm{B}}T}{\sqrt{(m_0 c^2)^2 + c^2 p^2}} \right), \end{aligned} \tag{10}$$

$$\begin{aligned} \hbar(\omega - \omega_o) &= \tfrac{15}{\pi^4} \hbar\omega_0 \left(\tfrac{k_{\mathrm{B}}T}{\hbar\omega_0}\right) \int_0^\infty \mathrm{d}x \tfrac{x^3}{e^x - 1} \sqrt{\left(\tfrac{\hbar\omega_0}{k_{\mathrm{B}}T}\right)^2 + x^2} - \tfrac{15}{\pi^4} \hbar\omega_0 \int_0^\infty \mathrm{d}x \tfrac{x^3}{e^x - 1} \\ &= \frac{C}{\pi^4} \hbar\omega_0 \int_0^\infty \mathrm{d}x \frac{x^3}{e^x - 1} \left(\sqrt{1 + \left(\frac{k_{\mathrm{B}}T}{\hbar\omega_0}\right)^2 x^2} - 1 \right) = \hbar\omega_0 \left(\mathcal{I}_{1/2} \left(\frac{k_{\mathrm{B}}T}{\hbar\omega_0}\right) - 1 \right) \end{aligned}, \tag{11}$$

IOP Conf. Series: Journal of Physics: Conf. Series **865** (2017) 012013 doi:10.1088/1742-6596/865/1/012013

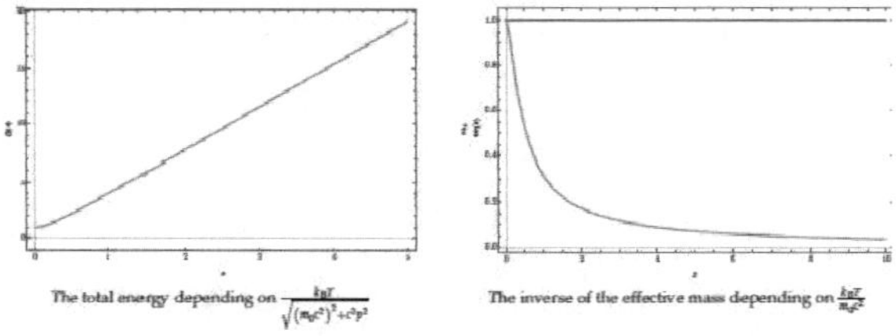

Figure 1. The total energy and effective mass depending on $k_B T$

and the inverse of the effective mass

$$\frac{1}{m^*} = \frac{15c^2}{\pi^4 (k_{\rm B}T)^4} \int\limits_0^\infty d\epsilon \frac{\epsilon^3}{e^{\frac{\epsilon}{k_{\rm B}T}} - 1} \frac{1}{\sqrt{(m_0c^2)^2 + \varepsilon^2}} = \frac{1}{m_0} I_{-\frac{1}{2}}(\frac{k_B T}{m_0 c^2}), \tag{12}$$

respectively, where the auxiliary function $I_\gamma(\alpha)$ is analytically defined by an infinite series of α^2 as

$$\begin{aligned} I_\gamma(\alpha) &= \tfrac{15}{\pi^4} \int\limits_0^\infty dx (1+\alpha^2 x^2)^\gamma \tfrac{x^3}{e^x - 1} \\ &= \tfrac{15}{\pi^4} \sum\limits_0^\infty \tfrac{(2i+3)!}{i!} \gamma(\gamma-1)\text{.......}(\gamma - i + 1) Li_{2i+4}(1) \alpha^{2i} \end{aligned} \tag{13}$$

It is obvious from the equation (12) that non–zero average value of fluctuation generates non–zero mass of a massless particle

$$m^* c^2 = (\frac{15}{\pi^4}(2\zeta(3)))^{-1}(k_B T) = 2.70118(k_B T), \tag{14}$$

2.4. The Frequency Shift and the de Broglie Wavelength

In contrast to the particle formalism, in the wave formalism the frequency shift effect is obtained

$$\frac{\omega}{\omega_0} = I_{\frac{1}{2}}(\frac{k_B T}{\hbar\omega}) \tag{15}$$

As it is illustrated in the left panel of the above figure, function $I_{\frac{1}{2}}(\frac{k_B T}{\hbar\omega})$ is always greater than one.

This means that under the influence of black body fluctuation the frequency of initial wave increases,or in other words, the ultraviolet shift is observable. The de Broglie wavelength of wave propagating in the black body radiation is changed and determined by the expression

$$\frac{\lambda}{\lambda_0} = I_{\frac{1}{2}}(\frac{k_B T}{\hbar\omega}) \tag{16}$$

Also, as it is illustrated in the left panel of the above figure, the ratio$\frac{\lambda}{\lambda_0}$ gets maximum at zero temperature $T = 0$, and then it decreases while temperature is increasing.

3. Conclusion

In conclusion, by considering the probability density functions of background fluctuations modeled by black body radiation, the change of effective mass of a moving massive and massless particles has been studied. In the case of massless particle, the black body fluctuation generates non–zero mass which is linearly depending on background temperature. In the case of massive particle, the dependence of effective mass on the ratio of mass and temperature of background fluctuations, $k_B T$, has been analytically found. Furthermore, the frequency shift and the de Broglie wavelength are also analytically obtained.

Acknowledgments

This research is funded by Vietnam National Foundation for Science and Technology Development (NAFOSTED) under grant number 103.01-2015.42.

References

[1] Black Body in *Wikipedia* 2017 https://en.wikipedia.org/wiki/Black_body
[2] Einstein A 1905 Zur Elektrodynamik bewegter Körper *Annalen der Physik* (in German) **322**(10) 891–921
[3] Einstein A 1906 Über eine Methode zur Bestimmung des Verhältnisses der transversalen und longitudinalen Masse des Elektrons *Annalen der Physik* (in German) **21**(13) 583–586
[4] John Roche 2005 What is mass? *European Journal of Physics 2005* **26**(2) 225
[5] Okun L B 1989 The Concept of Mass *Physics Today* **42**(6) 31–36
[6] Landau L D 1946 On the vibration of the electronic plasma *J. Phys. USSR* **10** 25
[7] Landau L D and Lifshitz E M 1984 *Electrodynamics of Continuous Media* (Pergamon Press)
[8] Andreev A A 2000 *An Introduction to Hot Laser Plasma Physics* (Huntington, New York: Nova Science Publishers, Inc.)
[9] Lyon A 2014 Why are Normal Distributions Normal? *The British Journal for the Philosophy of Science* **65**(3) 621–649
[10] Park Sung Y and Bera Anil K 2009 Maximum entropy autoregressive conditional heteroskedasticity model *Journal of Econometrics* **150** 219–230
[11] Dirac P 1958 *The Principles of Quantum Mechanics* 4th ed. (Oxford at the Clarendon Press)
[12] Gauss C F 1857 *Theory of the Motion of the Heavenly Bodies Moving about the Sun in Conic Sections: A translation of Gauss's "Theoria motus," with an Appendix by Charles Henry Davis* (Boston: Little, Brown and Company)

IOP Conf. Series: Journal of Physics: Conf. Series **865** (2017) 012014 doi:10.1088/1742-6596/865/1/012014

The spatially anisotropic triangular lattice antiferromagnet: Popov-Fedotov method

Pham Thi Thanh Nga[1], Phan Thu Trang[2] and Nguyen Toan Thang[3]

[1] Thuyloi University, 175 Tay Son, Hanoi, Vietnam
[2] Hanoi National University of Education,136 Xuan Thuy, Hanoi, Vietnam
[3] Institute of Physics, 10 Dao Tan, Hanoi, Vietnam

E-mail: nga_ptt@tlu.edu.vn

Abstract. We present an analysis of the antiferromagnetic Heisenberg model on an triangular lattice with spatially anisotropic J_1-J_2 exchange interactions. We apply the Popov-Fedotov method based on introducing an imaginary valued chemical potential to enforce the auxiliary fermion constraint exactly. The staggered magnetization, magnon spectra, free energy are computed in one loop approximation and compared using two different constraints: exact and on average. In the limit of zero temperature the results are identical, whereas at higher temperature significant differences are found. The comparisons with the results obtained by other methods are discussed.

1. Introduction

Frustrated quantum antiferromagnets have been much studied in resent years. One of the challenges is to understanding the interplay between quantum fluctuations and frustration [1, 2]. In this context, the triangular lattice Heisenberg antiferromagnet is a such magnet. The isotropic $S = 1/2$ quantum Heisenberg antiferromagnet (HAF) on the triangular lattice has been studied for the last few decades by analytical [3, 4] as well as numerical [5 - 7] methods. It is widely believed that the classical noncollinear 120° spin structure is stable in the quantum case with the strong reduction of staggered moment [4]. However, there exists no good material candidate for the pure isotropic triangular lattice HAF. The real magnets have considerable spatially in-plane anisotropies [8], in particular for Cs_2CuCl_4 [9, 10] and Cs_2CuBr_4 [10, 11] compounds. There are several analytical [12-15] and numerical [16, 17] investigations of the anisotropic triangular model.

The goal of this report is to study the spin $S = 1/2$ Heisenberg antiferromagnet on anisotropic triangular lattice, applying Popov-Fedotov formalism [18] for the semi-fermionic spin representation. The use of different representations of the spin operators in terms of the auxiliary Fermi or Bose operators is one basic approach to the investigation of spin systems. However, the representation of spins as the bilinear canonical operators enlarges the dimensionality of Hilbert space in which these operators are acting. Thus, one must impose the constraint on the auxiliary operators to remove the unphysical states from the consideration. The simplest way to cure the constraint problem is the replacement of the local by a so-called global constraint, where the number of auxiliary particles is fixed only in the average over all sites. Popov-Fedotov proposed a new method when the local constraint is taken into account exactly by introducing

IOP Conf. Series: Journal of Physics: Conf. Series **865** (2017) 012014 doi:10.1088/1742-6596/865/1/012014

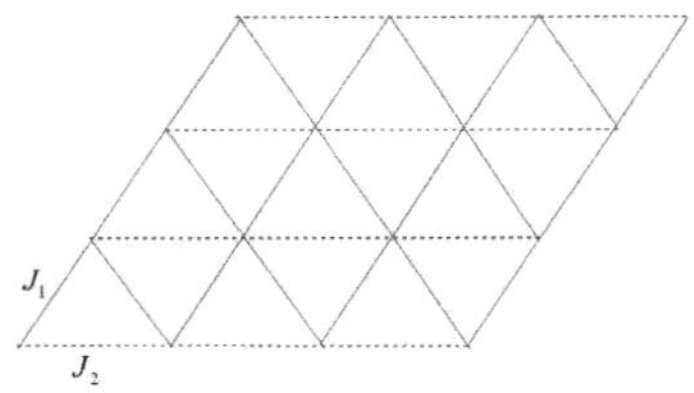

Figure 1. Anisotropic triangular lattice

the proper imaginary chemical potential for spin $S = 1/2$ and $S = 1$. Later, Veits at al have extended the Popov-Fedotov trick for arbitrary spin [19]. The Popov-Fedotov formalism then has been applied for the various spin-1/2 Heisenberg models [20-25].

In this paper we study the Heisenberg antiferromagnet on an anisotropic triangular lattice, following closely the calculations carried out in [25]. The plan of the paper is the following. In section II, we introduce the model and the classical phases. In the section III, we present the results for the sublattice magnetization, the free energy and the specific heat. In the last section we shall conclude with our results.

2. Model and classical state

The Hamiltonian of the Heisenberg antiferromagnet on an anisotropic triangular lattice is written as:

$$H = J_1 \sum_i \sum_{j=i+\vec{\delta}_{2,3}} \vec{S}_i.\vec{S}_j + J_2 \sum_i \sum_{k=i+\vec{\delta}_1} \vec{S}_i.\vec{S}_k \tag{1}$$

where $\vec{S}_i$ denotes the $S = 1$ spin vector operator. The first sum runs over all nearest - neighbour pair along the $x-$axis with the antiferromagnetic exchange interaction $J_1(J_1 > 0)$ and the second sum runs over all other nearest neighbour pairs (See Fig 1.) with the interaction $J_2(J_2 > 0)$.

The nearest-neighbour bond vectors are given by:

$$\vec{\delta}_1 = a(1,0); \vec{\delta}_2 = a\left(-\frac{1}{2}, \frac{\sqrt{3}}{2}\right); \vec{\delta}_3 = a\left(\frac{1}{2}, \frac{\sqrt{3}}{2}\right) \tag{2}$$

The bonds J_1 and J_2 are competing interations and lead to magnetic frustration. Note that in the limit $J_1 = 0$ the model reduces to decoupled chains, the limit $J_1 = J_2$ coresponds to the isotropic triangular lattice, while in the case $J_2 = 0$ the model becomes topologically equivalent to the square lattice. In two dimensional frustrated lattice described by Hamiltonian (1) the classical ground states have long range order, which are the collinear ordering and the spiral ordering [1, 2]. Both orderings are obtained in the following.

In the classical limit of $S = \infty$, we assume that the spins are planar in the plane Oxz and are described by some magnetic ordering vector $\vec{Q}$ as follows:

$$\vec{S}_i = S\left(\vec{u}\sin\vec{Q}\vec{r}_i + \vec{v}\cos\vec{Q}\vec{r}_i\right) \tag{3}$$

where $\vec{u}, \vec{v}$ are two orthonormal unit vectors in the plane Oxz. The vector $\vec{Q}$ defines the relative orientation of the spins on the lattice, namely an angle between the vectors $\vec{S}_i$ and $\vec{S}_j$ is given by:

$$\theta_{ij} = \theta_i - \theta_j = \vec{Q}\left(\vec{r}_i - \vec{r}_j\right) \tag{4}$$

Inserting $\vec{S_i}$ into Hamiltonian (1) we get the classical energy in terms of the ordering vector $\vec{Q}$ as folows:

$$E_{cl} = \frac{1}{2} N S^2 J\left(\vec{Q}\right) \tag{5}$$

where N is site number of the lattice and $J\left(\vec{Q}\right)$ is the Fourier transform of the exchange integral:

$$J\left(\vec{Q}\right) = \sum_{\vec{\delta}_1} J_{\vec{\delta}_1} \cos\left(\vec{Q}\vec{\delta}_1\right) = 2J_1\left(\alpha \mathrm{cos} \mathrm{Q}_x + 2\mathrm{cos}\frac{Q_x}{2}\cos\frac{\sqrt{3}}{2}\mathrm{Q}_y\right) \tag{6}$$

where $\alpha = \frac{J_2}{J_1}$ and we let the lattice $a = 1$.

By minimizing the classical energy (5) with respect to the vector $\vec{Q}$, we find two kinds ordered phases:

i) Collinear Neel state is characterized by $\vec{Q} = \left(0, \frac{2\pi}{\sqrt{3}}\right)$ for the region $0 \le \alpha \le \frac{1}{2}$.

ii) Incommensurate spiral state with $\vec{Q} = \left(2\mathrm{cos}^{-1}\left(-\frac{1}{2\alpha}\right), 0\right)$,where $\theta = \arccos\left(-\frac{1}{2\alpha}\right)$. For $\alpha = 1$ we have a triangular lattice which corresponds to the 120^ospin structure.

In the following we study the influence of the fluctuations on the classical state by means of Popov-Fedotov functional integral method, applying the calculations done in [25].

3. The resuls

3.1. Mean-field approximation

Working in the local coordinate system and parametrizing the classical state by ordering vector, one is able to express in a unique form the results for both the Neel and the spiral state. We introduce notations for $\lambda, \gamma_x\left(\vec{p}\right), \gamma_y\left(\vec{p}\right), \gamma_\omega\left(\vec{p}\right)$:

a) For Neel state: $0 \le \alpha \le \frac{1}{2}$,

$$\lambda = 2(2-\alpha)J_1 \tag{7}$$

$$\begin{cases} \gamma_x\left(\vec{p}\right) = \frac{1}{\alpha - 2}\left(\alpha\cos p_x - 2\cos\frac{p_x}{2}\cos\frac{\sqrt{3}}{2}p_y\right) \\ \gamma_y\left(\vec{p}\right) = \frac{1}{\alpha - 2}\left(\alpha\cos p_x + 2\cos\frac{p_x}{2}\cos\frac{\sqrt{3}}{2}p_y\right) \\ \gamma_\omega\left(\vec{p}\right) = 0 \end{cases} \tag{8}$$

b) For the spiral state: $\alpha > \frac{1}{2}$,

$$\lambda = 2J_1(\alpha + \frac{1}{2\alpha}) = \frac{J_1}{\alpha}(2\alpha^2 + 1) \tag{9}$$

$$\begin{cases} \gamma_x\left(\vec{p}\right) = \frac{1}{2\alpha^2+1}\left[\left(2\alpha^2 - 1\right)\cos p_x + 2\cos\frac{p_x}{2}\cos\frac{\sqrt{3}}{2}p_y\right] \\ \gamma_y\left(\vec{p}\right) = -\frac{2\alpha}{2\alpha^2+1}\left(\alpha\cos p_x + 2\cos\frac{p_x}{2}\cos\frac{\sqrt{3}}{2}p_y\right) \\ \gamma_\omega\left(\vec{p}\right) = -\frac{\sqrt{4\alpha^2-1}}{2\alpha^2+1}\left(\sin p_x - 2\sin\frac{p_x}{2}\cos\frac{\sqrt{3}}{2}p_y\right) \end{cases} \tag{10}$$

The sublattice magnetization per site in the mean field approximation reads:

$$m_o = \frac{1}{2}\tanh\left(\beta\lambda m_o\right) \tag{11}$$

In the case of average constraint, instead of equation (11) one has:

$$\tilde{m}_o = \frac{1}{2}\tanh\left(\frac{\beta\lambda\tilde{m}_o}{2}\right) \tag{12}$$

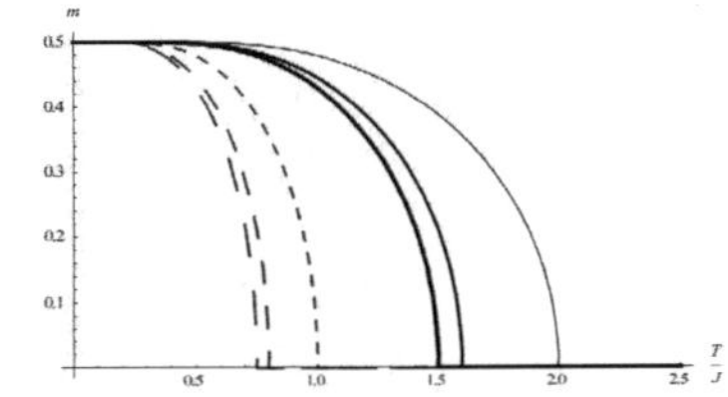

Figure 2. Temperature dependence of mean field magnetization m_o (for exact constraint: full line, from right to left) and $\tilde{m}_o$ (for: average constraint: dashed line, from right to left) for Neel state with $\alpha = 0, \alpha = 0.4, \alpha = 0.5$.

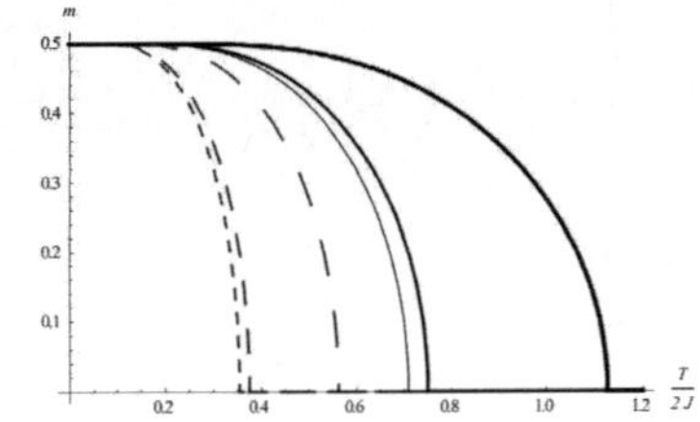

Figure 3. Temperature dependence of mean field magnetization m_o (for exact constraint: full line, from left to right) and $\tilde{m}_o$ (for: average constraint: dashed line, from left to right) for incommensurate spiral phase with $\alpha = 0.75, \alpha = 1, \alpha = 2$.

Accordingly, the equations result in different critical temperature:

$$\begin{array}{l} \text{exact-constraint} : T_C = \frac{\lambda}{2} \\ \text{average-constraint} : T_{Co} = \frac{\lambda}{4} \end{array} \tag{13}$$

From (13) we see the critical temperature for the case of exact constraint is higher than for average constraint is higher than for average constraint $\frac{T_C}{T_{Co}} = 2$. This is due to thermal fluctuations into unphysical spinless states when constraint is treated globaly, which reduce the magnetic moment. Howerver, at zero temperature both constraint methods give the same result. In Fig 2 and Fig 3 we plot the temperature dependence of the mean field sublattice magnetization for Neel and Spiral states, respectively. Fig 2 and Fig 3 show the difference of the results for the case of exact constraint and average one.

The mean field free energy per site can be expressed in term of the magnetization $m_o$as follows:

$$\frac{F_{MF}}{N} = \frac{1}{2}\lambda m_o^2 - \frac{1}{\beta}\ln\left(2\cosh\frac{\beta\lambda \mathrm{m}_o}{2}\right) \tag{14}$$

In the case of average projection instead of (14) one has:

$$\frac{F_{MF}}{N} = \frac{1}{2}\lambda\tilde{m}_o^2 - \frac{2}{\beta}\ln\left(2\cosh\frac{\beta\lambda\tilde{m}_o}{2}\right) \tag{15}$$

Again, at $T = 0K$, the Eqs. (14) and (15) lead to the same result.

3.2. Fluctuation contributions

In one loop approximation we obtain the fluctuation contributions to the longitudinal δm_{zz} and transerve δm_{+-} parts for the magnetization as follows:

$$\delta m^{zz} = -\frac{1}{2N}\sum_{\vec{p}}\frac{B_o}{A_o} \tag{16}$$

$$\delta m^{+-} = -\frac{1}{2N}\sum_{\vec{p}}\coth\frac{\beta E(\vec{p})}{2}\left[\Delta m_o\lambda\omega(\vec{p}) + \frac{1-\frac{1}{2}\frac{\gamma_x(\vec{p})+\gamma_y(\vec{p})}{6J}}{\omega(\vec{p})}\right] + \frac{1}{2}\coth\frac{\beta\lambda m_o}{2}\left(1+\lambda\Delta m_o\right) \tag{17}$$

where:

$$A_o = 1 + \beta\lambda C_o\left(\gamma_x(\vec{p}) + \frac{\gamma_w^2(\vec{p})}{\gamma_x(\vec{p})-1}\right) \tag{18}$$

$$B_o = \beta\lambda\left(\gamma_x(\vec{p}) + \frac{\gamma_w^2(\vec{p})}{\gamma_x(\vec{p})-1}D_o + \frac{\gamma_w^2(\vec{p})}{\left(\gamma_x(\vec{p})-1\right)^2}\frac{C_o}{m_o}\right) \tag{19}$$

C_o and D_o is given by:

$$C_o = -\frac{1}{4}\left(1-m_o^2\right) \tag{20}$$

$$D_o = m_o\Delta m_o \tag{21}$$

$$\Delta m_o = -\frac{\beta C_o}{1+\beta\lambda C_o} \tag{22}$$

The magnon energy reads:

$$\begin{cases} E(\vec{p}) = \lambda m_0\omega(\vec{p}) \\ \omega^2(\vec{p}) = \left(1-\gamma_x(\vec{p})\right)\left(1-\gamma_y(\vec{p})\right) \end{cases} \tag{23}$$

Note that in the equations (16), (17) an in the following the sums run over the whole Brillouin zone. The contributions of the longitudinal and transverse fluctuations to the free per site are given by:

$$\delta F_{zz} = \frac{1}{2\beta}\sum_{\vec{p}}\ln A_0 \tag{24}$$

$$\delta F_{+-} = \frac{1}{\beta}\sum_{\vec{p}}\ln\frac{\mathrm{sh}\frac{\beta E(\vec{p})}{2}}{\mathrm{sh}\frac{\beta\lambda m_0}{2}} \tag{25}$$

Where the single occupancy condition is disregarded, on the equations from (23) to (25) the following replacement should be done:

$$m_o \to \tilde{m}_o \tag{26}$$

It is easy to check that at zero temperature limit it does not matter whether the constraint is treated exactly or on the average the magnetization and the free energy are the same for both cases of the constraint conditions. Taking the limit of zero temperature $T \to 0K$ for the above equation we get the same sublattice magnetization and the ground state energy obtained in linear spin wave approximation by mean of Holstein- Primakov representation [12-15].

On the contrary, at finite temperature the exact constraint lead to the different values of the thermodynamical quantities, e.g., the magnetization, the internal energy, the specific heat...Fig 4 and Fig 5 show the difference in the finite temperature properties of the two methods of treating the constraint conditions for both ordered states.

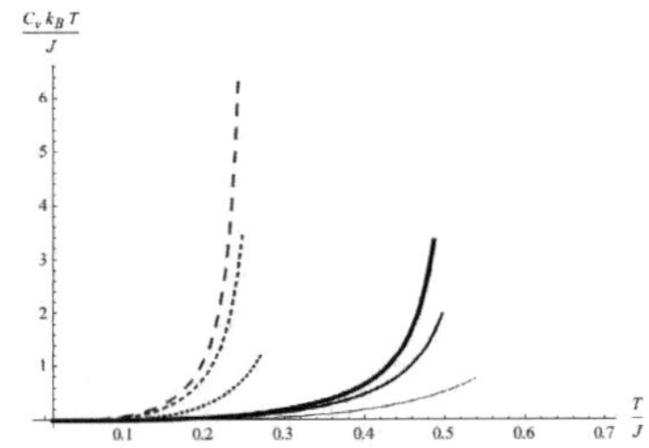

Figure 4. Temperature dependence of specific heat (for exact constraint: full line, from right to left and for average constraint: dashed line, from right to left) for collinear phase with $\alpha = 0, \alpha = 0.4, \alpha = 0.5$.

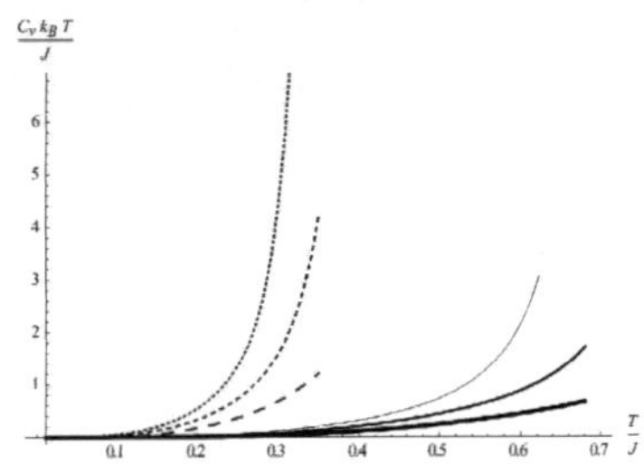

Figure 5. Temperature dependence of specific heat (for exact constraint: full line, from left to right and for average constraint: dashed line, from left to right) for incommensurate spiral phase with $\alpha = 0.75, \alpha = 1, \alpha = 2$.

4. Conclusion

We get the ground state properties of the spatially anisotropic triangular magnets consistent with the results obtained by mean of other linear spin wave theories [12-15]. This confirms the fact that at the zero temperature the Popov-Fedotov approach does not improve the results when the constraint on slave particle is treated approximately.

At the finite temperatures the exact projection onto physical Hilbert space gives significant differences in comparison with average projection. This is due to thermal fluctuation of slave fermions into unphysical spinless state. At the parameter range of interest, our results agree with results of numerical investigstions [16, 17].

It would be interesting to see how to extend the method used here to study the spin-liquid region and for spin quantum number $S > 1/2$.

Acknowledgment

This work was financially supported by the National Foundation for Science and Technology Development (NAFOSTED) of Vietnam under Grant No 103.01- 2014.23.

References

[1] Richter J, Schalenburg J, and Honecker A 2004, in *"Quantum magnetism"*, Lecture notes in Physics 645, ed. by Schollwork U, Richter J, Farnell D J J and Bishop R F, Springer, Berlin, p.85
[2] Misguick G and Lhuillier C 2004, in *Frustrated Spin Systems*,ed. by H. T. Diep, World Scientific Singapore, p. 299
[3] Jolicocur Th and Le Guillou J C 1989, *Phys. Rev.* B **40** 2727
[4] Chernyshev A L and Zhitomirsky M E 2009, *Phys. Rev.* B **79** 144416
[5] Bernu B, Lecheminant P, Lhuillier C and Pierre L 1994,*Phys. Rev.* B **50** 10048
[6] Wite S R and Chernyshev A L 2007, *Phys. Rev. lett.* **99** 127004

[7] Kulagin S A, Prokof'ev N, Starykh O A, Svistunov B and Varney C N 2013, *Phys. Rev.* B **87** 024407
[8] Nakatsuji S, Nambu Y, and Onada S 2010, *J. Phys. Soc. Jpn.* **79** 01103
[9] Coldea R, Tennant D A, and Tylczynski Z 2003, *Phys. Rev.* B **68** 13442
[10] Zvyagin S et al 2014, *Phys. Rev. lett* **112** 077206
[11] Ono T et al 2003, *Phys. Rev.* B **67** 104431
[12] Trumper A E 1999, *Phys. Rev.* B **60** 2987
[13] Morino J, Mckenzie R H, Marston J B and Chung C H 1999, *J. Phys: Condens. Matter,* **11** 2965
[14] Bhaumik U and Taraphder A 2006, *J. Phys. Condens. Matter,* **18** 8251
[15] Schmidt B and Thalmeier P 2014, *Phys. Rev.* B **89** 184402
[16] Zheng W, Singh R R P, Mckenzie R H and Coldea R 2005, *Phys. Rev.* B **71** 134422
[17] Bishop R F, Li P H Y, Farrell D J J, Campell C F 2009, *Phys. Rev.* B **79** 174405
[18] Popov V N and Fedotov S A 1988, *Sov. Phys. JETP* **67** 535
[19] Veits O et al 1994, *J. France* **4** 493
[20] Tejima S and Oguchi A 1995, *J. Phys. Soc. Jpn.* **64** 4923
[21] Azakov S, Dilaver M and Oztas 2000, *Int. J. Mod. Phys.* B **14** 13
[22] Dillenschneider R and Richert J 2006, *Eur. Phys. J.* B **49** 187
[23] Kicelev M N, Feldmann H and Opperman R 2002, *Phys. Rev.* B **65** 184410
[24] Kicelev M N 2006, *Int. J. Mod. Phys.* B **20**, 381.
[25] Pham Thi Thanh Nga and Nguyen Toan Thang 2012, *Comm. in Phys.* **22** 33; *Comm. in Phys.* **22** 383 Erratum.

Magnetic order of spin $S = 1$ antiferromagnetic quantum Heisenberg systems on a Bravais lattice: exact local constraint

Pham Thi Thanh Nga[1] **and Nguyen Toan Thang**[2]

[1] Thuyloi University, 175 Tay Son, Dong Da, Hanoi, Vietnam
[2] Institute of Physics, 10 Dao Tan, Ba Dinh, Hanoi, Vietnam

E-mail: nga_ptt@tlu.edu.vn

Abstract. We apply the Popov-Fedotov formalism for investigating magnetic properties of the spin $S = 1$ Heisenberg antiferomagnetic (HAF) on a Bravais lattice. Mapping spin-1 lattice systems on three-component auxiliary fermions with imaginary chemical potential and transforming to a local coordinate system allow us to express a sublattice magnetization, free energy and other thermodynamical quantities in unique forms for different lattice structures in various magnetically ordered phases. We compare them with the results obtained when the local constraint is disregarded. A comparison with the case of $S = 1/2$ is also discussed.

1. Introduction

The fact that spin operators satisfy the non-canonical commutation relations poses a great difficulty to study theoretically quantum spin systems, because it leads to the absence of Wick theorem [1], so it is impossible to apply a standard perturbation technique directly for spin operators. One of the ways to resolve this problem is to represent the spin operators in terms of the auxiliary Fermi or Bose-operators [1]. However, any representation of spin operators as a bilinear combination of auxiliary operators makes the Hilbert space for spin operators larger. As a result, one must remove the spurious unphysical states by imposing some constraint on the auxiliary operators for each lattice site. In general it is very difficult to treat the local constraint exactly. One simple common approximation is to replace the local constraint by a so-called global constraint, where the number of auxiliary particles is taken to be on average for the whole lattice system. Popov-Fedotov proposed a method for quantum spin systems with $S = 1/2$ and $S = 1$ free of the local constraint problem [2], introducing some proper imaginary chemical potential. They demonstrated that if the usual Fermionic Matsubara frequencies $\omega = \frac{2\pi}{\beta}\left(n + \frac{1}{2}\right)$ are replaced by $\omega = \frac{2\pi}{\beta}\left(n + \frac{1}{4}\right)$ for $S = 1/2$ and by $\omega = \frac{2\pi}{\beta}\left(n + \frac{1}{3}\right)$ for $S = 1$ then the contributions of the unphysical states cancel each other and give no contribution to the partition function. Later, the extention of the Popov-Fedotov formalism for arbitrary spin was done by Veits et al [3]. In fact, this paper is basically of a methodologically nature. The concrete calculations then have been performed for various spin systems with $S = 1/2$: three dimensional ferromagnetic Heisenberg model [4], Heisenberg antiferromagnet on a d-dimensional hypercubical [5, 6] and a triangular [7] lattice. However, many real magnets have spin quantum number $S = 1$, in particular the well studied experimentally material $NiGaS_4$. Besides, the spin quantum number

S can play a considerable role in the various magnetic phenomena. A well-known example of this kind is the Haldane phase for $S = 1/2$ one dimensional Heisenberg model, which is absent in his $S = 1/2$ counter parts [8]. It is thus of great interest to apply the Popov-Fedotov formalism to spin systems with $S = 1$. In this report, we derive the general expressions of a sublattice magnetization, free energy and other thermodynamical quantities in unique forms for a Bravais lattice in a magnetically ordered state. The plan of the report is the following. In section II we introduce the model and the formalism. The mean field approximation is considered in section III. The fluctuations around a mean-field state are investigated in section IV. The last section is devoted to conclusion and discussions.

2. Formulation

We consider a general antiferromagnetic Heisenberg Hamiltonian on a Bravais lattice given by:

$$H_S = \sum_{ij} J_{ij} \vec{S}_i.\vec{S}_j \tag{1}$$

where $\vec{S}_i$ denotes the $S = 1$ spin vector operator. The interaction between the sites are antiferromagnetic ($J_{ij} > 0$). In the general frustrated lattice described by Hamiltionan (1) the classical ground states have long range order, which may be parametrized by some magnetic ordering vector $\vec{Q}$ [9]. We assume that the spins are planar in the plane Oxz and are described as follows [9]:

$$\vec{S}_i = S\left(\vec{u}\sin\vec{Q}\vec{r}_i + \vec{v}\cos\vec{Q}\vec{r}_i\right) \tag{2}$$

where $\vec{u}, \vec{v}$ are two orthonormal unit vectors in the plane Oxz. The vector $\vec{Q}$ defines the relative orientation of the spins on the lattice, namely an angle between the vectors $\vec{S}_i$ and $\vec{S}_j$ is given by:

$$\theta_{ij} = \theta_i - \theta_j = \vec{Q}\left(\vec{r}_i - \vec{r}_j\right) \tag{3}$$

Inserting $\vec{S}_i$ into Hamiltonian (1) we get the classical energy in terms of the ordering vector $\vec{Q}$ as folows:

$$E_{cl} = \frac{1}{2} N S^2 J\left(\vec{Q}\right) \tag{4}$$

where N is site number of the lattice and $J\left(\vec{Q}\right)$ is the Fourier transform of the exchange integral:

$$J\left(\vec{Q}\right) = \frac{1}{N}\sum_{ij} J_{ij} e^{-i\vec{Q}(\vec{r}_i - \vec{r}_j)} \tag{5}$$

The magnetic ordering vector $\vec{Q}$ can be obtained by minimizing (5) with respect to $\vec{Q}$, i.e by setting $\frac{\partial J(\vec{Q})}{\partial Q_\alpha} = 0, \alpha =$ x, z.

In principle, there may exist several ordering vector $\vec{Q}$, minimizing the two-variable function $J\left(\vec{Q}\right)$. In order to incorporate the fluctuations in a unique way for all possible order phases, we folow Miyake [10], introducing a local coordinate system on each site i, where the local z axis is oriented along its classical direction. This transformation reads:

$$\begin{cases} S_i^z = S_i^{z'}\cos\theta_i - S_i^{x'}\sin\theta_i \\ S_i^x = S_i^{z'}\sin\theta_i - S_i^{x'}\cos\theta_i \\ S_i^y = S_i^{y'}. \end{cases} \tag{6}$$

IOP Conf. Series: Journal of Physics: Conf. Series **865** (2017) 012015 doi:10.1088/1742-6596/865/1/012015

Substituting (6) in (1), we get the following Hamiltonian, which is anisotropic on the spin space:

$$H_S = -\frac{1}{2} \sum_{\alpha\beta} J_{\text{ij}}^{\alpha\beta} S_i^\alpha S_j^\beta \tag{7}$$

where:

$$\begin{cases} J_{ij}^{xx} = J_{ij}^{zz} \equiv X_{ij} = -J_{ij}\cos\theta_{ij} \\ J_{ij}^{yy} \equiv Y_{ij} = -J_{ij} \\ J_{ij}^{zx} = -J_{ij}^{xz} \equiv W_{ij} = -J_{ij}\sin\theta_{ij} \\ J_{ij}^{xy} = J_{ij}^{yx} = J_{ij}^{yz} = J_{ij}^{zy} = 0 \end{cases} \tag{8}$$

The tranformation (6) allows us to introduce only one type of auxiliary particles for all lattice sites rather than one type for each magnetic sublattice. Following Popov-Fedotov [2], the $S = 1$ spin vector operators are written in terms of Fermi operators $f_{i\alpha}^+, f_{i\alpha}\,(\alpha = 1, 2, 3)$

$$\vec{S}_i = f_{i\alpha}^+ \tau_{\alpha\beta} f_{i\beta}\,(\alpha = 1, 2, 3) \tag{9}$$

where the $\vec{\tau}_{\alpha\beta}$ vector components are $S = 1$ $(3x3)$ matrices :

$$S_x = \frac{1}{\sqrt{2}} \begin{pmatrix} 0 & 1 & 0 \\ i & 0 & 1 \\ 0 & i & 0 \end{pmatrix}; S_y = \frac{1}{\sqrt{2}} = \begin{pmatrix} 0 & 1 & 0 \\ -i & 0 & 1 \\ 0 & -i & 0 \end{pmatrix}; S_z = \begin{pmatrix} 1 & 0 & 0 \\ 0 & 0 & 0 \\ 0 & 0 & -1 \end{pmatrix} \tag{10}$$

Here and in the following we put $\hbar = 1$. The cost of the representation (9) is the extention of the Hilbert space into unphysical sectors, which have to be eliminated by imposing a constraint for each site i:

$$\sum_\alpha f_{i\alpha}^\dagger f_{i\alpha} = 1 \tag{11}$$

As it was shown by Popov-Fedotov [2], this can be done by means of introducing the projection operator $\hat{\mathrm{P}} = e^{i\mu\hat{N}_F}$ with $\hat{N}_F = \sum_{i\alpha} f_{i\alpha}^+ f_{i\alpha}$ being number operator and $\mu = i\frac{\pi}{3\beta}, \beta = k_B T$ being the imaginary chemical potential. The partition of a spin system with Hamiltonian H_s can be expressed in term of Fermi operators as follows:

$$Z = \left(\frac{i}{\sqrt{3}}\right)^N Tr\left[e^{-\beta(\hat{H}_F - \mu\hat{N}_F}\right] \tag{12}$$

here $\hat{H}_F$ is Hamiltonian H_s in the fermionic representation (9). The appearence of the imaginary chemical potential in (12) results in that the corresponding Matsubara frequencies for Popov-Fedotov fermions read as:

$$\omega_n = \frac{2\pi}{\beta}\left(n + \frac{1}{3}\right) \tag{13}$$

The partition fuction Z in (12) can be treated following the standard fuctional integral formalism similar to the case of $S = 1/2$ [6, 7]. The main differences are that instead of Pauli (2x2) matrices and Matsubara frequencies $\mu = \frac{2\pi}{\beta}\left(n + \frac{1}{4}\right)$ we have to deal with the spin $S = 1$ (3x3) matrices (10) and Matsubara frequencies given by (13). The calculations are lenghthy but straightforward, so we only list here the main steps. Firstly, we express the partition fuction as path integral over Grassmann variables $\bar{\eta}_{i\alpha},\eta_{i\alpha}$ and transform the Heisenberg Hamiltonian (7) into bilinear fermionic expression by introducing a Hubbard Stratonovich decoupling:

$$\begin{cases} Z = \frac{i}{(\sqrt{3})^N} \frac{1}{Z_0} \int \prod_i D\left[\overrightarrow{\varphi_i}\right] \int D\left[\bar{\eta}_{i\alpha}, \eta_{i\alpha}\right] e^{-S\left[\overrightarrow{\varphi_i}, \bar{\eta}_{i\alpha}, \eta_{i\alpha}\right]} \\ Z_0 = \int \prod_i D\left[\overrightarrow{\varphi_i}\right] e^{-S_o\left[\overrightarrow{\varphi_i}\right]} \end{cases} \tag{14}$$

$$\begin{cases} S\left[\overrightarrow{\varphi_i},\bar{\eta}_{i\alpha},\eta_{i\alpha}\right] = S_o\left[\overrightarrow{\varphi_i}\right] + S_1\left[\overrightarrow{\varphi_i},\bar{\eta}_{i\alpha},\eta_{i\alpha}\right] \\ S_o\left[\overrightarrow{\varphi_i}\right] = \int\limits_0^\beta d\tau \left\{ \sum\limits_{ij\lambda\gamma} \left(J^{-1}\right)_{ij}^{\lambda\gamma} \varphi_i^\lambda(\tau)\varphi_j^\gamma(\tau) \right\} \\ S_1\left[\overrightarrow{\varphi_i},\bar{\eta}_{i\alpha},\eta_{i\alpha}\right] = \int\limits_0^\beta d\tau \left\{ \sum\limits_{i\alpha} \bar{\eta}_{i\alpha}(\tau)\left(\partial_\tau - \mu\right)\eta_{i\alpha}(\tau) + \sum\limits_{i,\alpha,\lambda} \bar{\eta}_{i\alpha}(\tau)\tau_{\alpha\beta}^\lambda \eta_{i\beta}(\tau)\varphi_i^\lambda(\tau) \right\} \end{cases} \tag{15}$$

where τ is an imaginary time and $\vec{\varphi}_i$ stands for the auxiliary Bose field, which plays the role of the staggered magnetization. Then we perform integration over the Grassmann variables to get:

$$Z = \left(\frac{i}{\sqrt{3}}\right)^N \frac{1}{Z_0} \int D\left[\overrightarrow{\varphi_i}\right] e^{-S_{eff}\left[\vec{\varphi}_i\right]} \tag{16}$$

where the effective action $S_{eff}\left[\vec{\varphi}_i\right]$ is given by:

$$S_{eff}\left[\vec{\varphi}_i\right] = S_o\left[\vec{\varphi}_i\right] - \ln\det\hat{K}_i \tag{17}$$

The matrix K_i in the frequency representation reads:

$$\hat{K}_i\left(\omega_1,\omega_2\right) = \left(-i\omega_1 - \frac{i\pi}{3\beta}\right)\delta_{\omega_1,\omega_2}\hat{I} + \frac{1}{2}\vec{\tau}\vec{\varphi}_i\left(\omega_1-\omega_2\right) \tag{18}$$

where $\hat{I}$ is (3x3) unit matrix.

Decomposing the matrix $\hat{K}$ into nonperturbation and perturbation parts:

$$\hat{K} = \hat{K}_o + \hat{M} \tag{19}$$

and expanding $Tr\left(\ln\hat{K}\right)$ in a Taylor series:

$$Tr\left(\ln\hat{K}\right) = Tr\left(\ln\hat{K}_o\right) + Tr\sum_{n=1}^{\infty}\frac{(-1)^{n+1}}{n}\left(\hat{K}_o^{-1}\hat{M}\right)^n \tag{20}$$

We can calculate the partition function in a given order of $\hat{M}$. The explicit forms of $\hat{K}_o$ and $\hat{M}$ depend on the way of decomposing the Hubbard Stratonovich auxilary field $\vec{\varphi}_i$. In the following we set:

$$\vec{\varphi}_i\left(\Omega\right) = \vec{\varphi}_{io}\left(\Omega=0\right) + \delta\vec{\varphi}\left(\Omega\right) \tag{21}$$

where $\vec{\varphi}_{io}$ is the mean field part and $\delta\vec{\varphi}\left(\Omega\right)$ is the fluctuation part of the auxilary field $\vec{\varphi}_i\left(\Omega\right)$.

Substituting (21) in (18), the expression (19) gives:

$$\begin{aligned} \hat{K}_o &= \left(i\omega_1\hat{I} + \vec{\tau}\vec{\varphi}_{io}\right)\delta_{\omega_1\omega_2}\delta_{\rm ij} \\ \hat{M} &= \vec{\tau}\delta\vec{\varphi}_i\left(\omega_1-\omega_2\right)\delta_{\rm ij} = \vec{\tau}\delta\vec{\varphi}_i\left(\Omega\right)\delta_{\rm ij} \end{aligned} \tag{22}$$

In order to calculate the sublattice magnetization we add a local magnetic field $\vec{B}_i$ to Hamiltonian (7) and let $\vec{B}_i = 0$ at the end of final calculations. Then at the one-loop approximation one derives the effective action as follows:

$$S_{eff} = S_{eff}^{(0)} + S_{eff}^{(1)} + S_{eff}^{(2)} \tag{23}$$

$$S_{eff}^{(0)} = \frac{1}{2}\sum_{ij\lambda\nu}\left(\varphi_{io}^\lambda + B_i^\lambda\right)\left(J^{-1}\right)_{ij}^{\lambda\nu}\left(\varphi_{jo}^\gamma + B_j^\nu\right) - \ln\det\hat{K}_o \tag{24}$$

IOP Conf. Series: Journal of Physics: Conf. Series **865** (2017) 012015 doi:10.1088/1742-6596/865/1/012015

$$S^{(1)}_{eff} = \frac{1}{2}\sum_{ij\lambda\nu}\left[\left(J^{-1}\right)^{\lambda\nu}_{ij}\left(\varphi^{\lambda}_{io}+B^{\lambda}_{i}\right)\delta\varphi^{\nu}_{j}(0)+\delta\varphi^{\lambda}_{i}(0)\left(\varphi^{\nu}_{jo}+B^{\nu}_{j}\right)\left(J^{-1}\right)^{\lambda\nu}_{ij}\right] - Tr\left(\hat{K}^{-1}_{o}\hat{M}\right) \tag{25}$$

$$S^{(2)}_{eff} = \frac{1}{2}\sum_{ij\lambda\gamma,\Omega}\left(J^{-1}\right)^{\gamma\lambda}_{ij}\delta\varphi^{\lambda}_{i}(-\Omega)\,\delta\varphi^{\gamma}_{j}(-\Omega)+\frac{1}{2}Tr\left(\hat{K}^{-1}_{o}\hat{M}\right)^{2} \tag{26}$$

The mean field φ^{α}_{io} is chosen by minimizing the effective action in the application of the least action principle.

Setting $\delta\varphi^{\pm}_{j}(\omega) = \delta\varphi^{x}_{j}(\omega) \pm \delta\varphi^{y}_{j}(\omega)$, the partition function can be decomposed into a product of three terms:

$$Z = Z_{MF}.Z_{zz}.Z_{+-} \tag{27}$$

where Z_{MF} is the mean field contribution, Z_{zz} and Z_{+-} are the fluctuation contributions for the longitudinal part $\delta\varphi^{z}(\omega)$ and the transverse parts $\delta\varphi^{+}(\omega), \delta\varphi^{-}(\omega)$. Consequently, the free energy may be written in the following way:

$$F = F_{MF} + \delta F_{zz} + \delta F_{+-} \tag{28}$$

3. Mean field approximation

Minimizing the efective action according the least action principle, we derive the mean field equations for the auxiliary fields φ^{α}_{io} :

$$\left.\frac{\delta S_{eff}}{\delta\varphi^{\alpha}_{i}}\right|_{\delta\varphi^{\alpha}_{i}=0} = 0 \tag{29}$$

From Eqs. (23) - (26) and (29) we get:

$$\sum_{\beta,j}\left(J^{-1}\right)^{\alpha\beta}_{ij}\left(\varphi^{\beta}_{jo}+B^{\beta}_{j}\right)\delta_{\Omega,0}+K^{\alpha}_{1i}(\Omega) = 0 \tag{30}$$

Where $K^{\alpha}_{1i}(\Omega)$ is defined as follows:

$$\sum_{i,\alpha,\Omega} K^{\alpha}_{1i}(\Omega)\,\delta\varphi^{\alpha}_{i}(\Omega) = -Tr\left(\hat{K}^{-1}_{0}\hat{M}\right) \tag{31}$$

From the definition of a magnetization per site i:

$$m^{\alpha}_{i} = -\frac{\partial F}{\partial B^{\alpha}_{i}} = -\frac{1}{\beta}\frac{\partial}{\partial B^{\alpha}_{i}}Z \tag{32}$$

one can find the relation between the local magnetization m^{α}_{i} and the auxiliary field φ^{α}_{i} . In particular, from (25) and (32), the mean field magnetization m^{α}_{io} related to the auxiliary field φ_{io} as follows:

$$\varphi^{\alpha}_{io} + B^{\alpha}_{i} = -\sum_{j\beta} m^{\beta}_{jo}J^{\beta\alpha}_{ji} \tag{33}$$

From (30) and (33), we obtain:

$$m^{\alpha}_{io} = K^{\alpha}_{1i} \tag{34}$$

Because we are working in the local coordinates with the classical magnetization pointing along the z-axis, thus, in the mean field approximation:

$$\begin{cases} m_{io}^{\alpha} = m_{io}\delta_{\alpha,z} \\ \varphi_{io}^{\alpha} = \varphi_{io}\delta_{\alpha,z} \end{cases} \tag{35}$$

Taking into account Eqs. (35), from (31) we get:

$$K_{1i}^{\alpha}(\Omega) = 2\sum_{i,\omega} \frac{\varphi_{io}}{(i\omega)^2 - \varphi_{io}^2}\delta_{\alpha,z}\delta_{\Omega,0} \tag{36}$$

For a Bravais lattice, all the sites are equivalent, so: $\varphi_{io} = \varphi_o; m_{io} = m_o$. The sum over the modified Matsubara frequencies can be carried out by the contour integration trick and gives:

$$K_{1i}^{\alpha}(\Omega) = \Delta(\varphi_o)\,\delta_{\alpha,z}\delta_{\Omega,0} \tag{37}$$

with $\Delta(\varphi_o)$ being:

$$\Delta(\varphi_o) = -\frac{2sh\beta\varphi_o}{1 + 2\cosh\beta\varphi_o} \tag{38}$$

Together with Eqs. (33) and (34), it leads to the following equation for the mean-field magnetization:

$$m_o = \Delta(\lambda m_o) \tag{39}$$

where:

$$\lambda = -J\left(\vec{Q}\right) \tag{40}$$

Note that in the case of everage projection $(\mu = 0)$, the sum over the fermionic Matsubara frequencies in (36) gives:

$$K_{1i}^{\alpha}(\Omega) = -th\frac{\beta\varphi_o}{2} \tag{41}$$

Then instead of Eq. (39) one has:

$$\tilde{m}_o = -th\left(\frac{\beta\lambda\tilde{m}_o}{2}\right) \tag{42}$$

Accordingly the equations (39) and (42) result in different critical temperature:

$$\begin{array}{l} \text{exact-constraint} : T_C = \frac{2\lambda}{3} \\ \text{average-constraint} : \tilde{T}_C = \frac{\lambda}{2} \end{array} \tag{43}$$

From (43) we obtains: $\frac{T_C}{\tilde{T}_C} = \frac{4}{3}$ for $S = 1$, while for the case of $S = 1/2$ one has $\frac{T_C}{\tilde{T}_C} = 2$ [5-7]. In both cases of $S = 1/2$ and $S = 1$ the critical temperature for the case of exact constraint is higher than for average one. This is due to thermal fluctuations into unphysical spinless states in the case of global constraint, which reduce the magnetic moment. In the case of spin $S = 1/2$ the number of physical states in the Fock space of the auxiliary fermion $N_{ph} = 2$ (one particle states), while two states are unphysical (the vacuum and the two-particle states): $N_{un} = 2$. Assuming $T_C \sim \frac{1}{N_{ph}}; \tilde{T}_C \sim \frac{1}{N_{ph}+N_{un}}$, it results in the ratio $\frac{T_C}{\tilde{T}_C} = 2$ for $S = 1/2$. For the case of spin $S = 1$, two states are unphysical (the vacuum and the three-particle states): $N_{un} = 2$. Due to the particle-hole symmetry of the spin Hamiltonian the three one-particle and three two-particle states are physical: $N_{ph} = 6$. Thus, one has for the $S = 1$ case: $\frac{T_C}{\tilde{T}_C} = \frac{4}{3}$. However, at zero temperature both constraint methods give the same result:

$$\lim_{T\to 0K} m_o = \lim_{T\to 0K} \tilde{m}_o = 1 \tag{44}$$

IOP Conf. Series: Journal of Physics: Conf. Series **865** (2017) 012015 doi:10.1088/1742-6596/865/1/012015

Free energy in mean-field approximation can be expressed in terms of the magnetization in the following way, letting the fictive magnetic field $B = 0$:

For exact constraint $(\mu = i\pi/3\beta)$:

$$F_{MF} = \frac{N\lambda m_o^2}{2} - \frac{2}{\beta}\ln\left(2\cosh\beta\lambda \mathrm{m}_o + 1\right) \tag{45}$$

For global constraint $(\mu = 0)$:

$$F_{MF} = \frac{N\lambda \tilde{m}_o^2}{2} - \frac{2}{\beta}\ln\left(2\cosh\frac{\beta\lambda \tilde{\mathrm{m}}_o}{2}\right) \tag{46}$$

4. Fluctuation contributions

Due to the Eq. (30), the first order fluctuation contributions vanish: $S^{(1)}{}_{eff} = 0$. The effective action in the second order of fluctuation (26) may be rewriten in the form:

$$S_{\mathrm{eff}}^{(2)} = \frac{1}{2}\sum_{\mathrm{ij}\alpha\beta} D_{ij}^{\alpha\beta}\left(\Omega\right)\delta\varphi_i^{\alpha}\left(-\Omega\right)\delta\varphi_j^{\beta}\left(\Omega\right) \tag{47}$$

where:

$$D_{ij}^{\alpha\beta}\left(\Omega\right) = \left(J^{-1}\right)_{ij}^{\alpha\beta}\left(\Omega\right) + K_{2ij}^{\alpha\beta}\left(\Omega\right) \tag{48}$$

In the local coordinates in the mean field approximation we have $\varphi_{io}^{\pm} = 0, \varphi_{io}^{z} = \varphi_o$ for every site i, therefore the kernel $K_{2ij}^{\alpha\beta}\left(\Omega\right)$ has only three nonzero components:

$$K_{2ij}^{zz}\left(\Omega\right) = \sum_{\omega}\sum_{k=-1}^{1}\frac{k^2}{\left(i\omega + k\varphi_o\right)\left(i\omega + \Omega + k\varphi_o\right)}\delta_{ij} \tag{49}$$

$$K_{2ij}^{+-}\left(\Omega\right) = \frac{1}{4}\sum_{\omega}\sum_{k=-1}^{0}\frac{2 - k(k+1)}{\left(i\omega + k\varphi_o\right)\left(i\omega + \Omega + (k+1)\,\varphi_o\right)}\delta_{ij} \tag{50}$$

$$K_{2ij}^{+-}\left(\Omega\right) = \frac{1}{4}\sum_{\omega}\sum_{k=0}^{1}\frac{2 - k(k-1)}{\left(i\omega + k\varphi_o\right)\left(i\omega + \Omega + (k-1)\,\varphi_o\right)}\delta_{ij} \tag{51}$$

Summing over the modified Matsubara frequencies, by mean of contour integration method,the above eqs give:

$$K_{2ij}^{zz}\left(\Omega\right) = -\frac{1}{3}\left(4 - 3\Delta^2\left(\varphi_o\right) - \sqrt{4 - 3\Delta^2\left(\varphi_o\right)}\right)\delta_{\Omega,0}\delta_{ij} \tag{52}$$

$$K_{2ij}^{+-}\left(\Omega\right) = \left(K_{2ij}^{+-}\right)^{*}\left(\Omega\right) = \frac{\beta}{2}\frac{\Delta\left(\varphi_o\right)}{\varphi_o + i\Omega}\delta_{ij} \tag{53}$$

The fluctuation corrections to the partition function Zfl are quadratic in the field variable $\delta\varphi_i^{+}, \delta\varphi_i^{-}, \delta\varphi_i^{z}$ and can be carried out:

$$Z_{fl} = \int D\left[\delta\vec{\varphi}\right]e^{-S_{\mathrm{eff}}\left[\delta\vec{\varphi}\right]} = \left[\det\hat{D}_{ij}\right]^{-1/2} = Z_o\left[\det\left(\hat{I} + \hat{J}_{ij}\hat{K}_{2\mathrm{ij}}\right)\right]^{-1/2} \tag{54}$$

IOP Conf. Series: Journal of Physics: Conf. Series **865** (2017) 012015 doi:10.1088/1742-6596/865/1/012015

The components of the exchange matrix $\hat{J}_{ij}$ may be given in the terms of X_{ij}, Y_{ij}, W_{ij} as follows:

$$\begin{cases} J_{ij}^{++} = J_{ij}^{--} = X_{ij} - Y_{ij} \\ J_{ij}^{+-} = J_{ij}^{-+} = X_{ij} + Y_{ij} \\ J_{ij}^{zz} = X_{ij} \\ J_{ij}^{+z} = J_{ij}^{-z} = -J_{ij}^{z+} = -J_{ij}^{z-} = -W_{ij} \end{cases} \tag{55}$$

After Fourier the components of the exchange interaction in (55) read:

$$\begin{cases} X(\vec{p}) = -\frac{1}{2}\left(J\left(\vec{p}-\vec{Q}\right) + J\left(\vec{p}+\vec{Q}\right)\right) \\ Y(\vec{p}) = -J(\vec{p}) \\ W(\vec{p}) = -\frac{1}{2}\left(J\left(\vec{p}+\vec{Q}\right) - J\left(\vec{p}-\vec{Q}\right)\right) \end{cases} \tag{56}$$

From the Eqs. (52) (54) and (56) we obtain the fluctuation contributions to the free energy:

$$\delta F_{zz} = \frac{1}{2\beta} \sum_{\vec{p}\in BZ} \ln A(\vec{p}) \tag{57}$$

$$\delta F_{+-} = \frac{1}{\beta} \sum_{\vec{p}\in BZ} \ln \frac{\mathrm{sh}\frac{\beta E(\vec{p})}{2}}{\mathrm{sh}\frac{\beta|\varphi_0|}{2}} \tag{58}$$

where:

$$A(\vec{p}) = 1 + X(\vec{p})\, K_2^{zz} + \frac{m_o K_2^{zz} W^2(\vec{p})}{\varphi_o + m_o X(\vec{p})} \tag{59}$$

$$E(\vec{p}) = \sqrt{(\varphi_o + m_o X(\vec{p}))\,(\varphi_o + m_o Y(\vec{p}))} \tag{60}$$

In the Eqs. (59) (60), the mean field auxiliary field φ_o in the presence of the fictive magnetic field B is related to the magnetization mo as follows:

$$\varphi_o = -B - \lambda\,(m_o + B\Delta m_o) \tag{61}$$

where:

$$\Delta m_o = -\frac{\beta K_2^{zz}(-\lambda m_o)}{1 + \beta\lambda K_2^{zz}(-\lambda m_o)} \tag{62}$$

Taking the derivative of the free energy (57) (58) with respect to the fictive magnetic field, we obtain the fluctuation contributions to the longitudinal δm_{zz} and transerve δm_{+-} parts for the magnetization as follows:

$$\delta m^{zz} = -\frac{1}{2N} \sum_{\vec{p}\in BZ} \frac{B_o(\vec{p})}{A_o(\vec{p})} \tag{63}$$

$$\delta m^{zz} = -\frac{1}{2N} \sum_{\vec{p}\in BZ} \frac{B_o(\vec{p})}{A_o(\vec{p})} \tag{64}$$

where:

$$A_o(\vec{p}) = 1 + \beta C_o \left(X(\vec{p}) + \frac{W^2(\vec{p})}{X(\vec{p}) - \lambda} \right) \tag{65}$$

$$B_o(\vec{p}) = \beta \left(X(\vec{p}) + \frac{W^2(\vec{p})}{X(\vec{p}) - \lambda} \right) D_o + \beta \frac{W^2(\vec{p}) C_o}{m_o(\lambda - X(\vec{p}))^2} \tag{66}$$

IOP Conf. Series: Journal of Physics: Conf. Series **865** (2017) 012015 doi:10.1088/1742-6596/865/1/012015

C_o and D_o is given by:

$$C_o = -\frac{1}{3}\left(4 - 3m_o^2 - \sqrt{4 - 3m_o^2}\right) \tag{67}$$

$$D_o = m_o \Delta m_o \left(2 - \frac{1}{\sqrt{4 - 3m_o^2}}\right) \tag{68}$$

The magnon energy reads:

$$\begin{cases} E(\vec{p}) = \lambda m_0 \omega(\vec{p}) \\ \omega^2(\vec{p}) = \left(1 - \frac{X(\vec{p})}{\lambda}\right)\left(1 - \frac{Y(\vec{p})}{\lambda}\right) \end{cases} \tag{69}$$

When the single occupancy condition is disregarded the following replacement should be done on the equations from (57) to (69):

$$\begin{cases} m_o \to \tilde{m}_o \\ K_2^{zz} \to \tilde{K}_2^{zz} = -\frac{1}{2}\left(1 - \tilde{m}_o^2\right) \end{cases} \tag{70}$$

Using the following relations:

$$U = \left(\frac{\partial \beta F}{\partial \beta}\right) \tag{71}$$

$$C_v = -\beta^2 \left(2\frac{\partial F}{\partial \beta} + \beta \frac{\partial^2 F}{\partial \beta^2}\right) \tag{72}$$

it is straightforward to derive explicit expressions for internal energy U and specific heat C_v from the Eqs. (45), (57), (58), putting the fictive magnetic field $B = 0$.

5. Conclusions and Discussions

In this paper we have obtained the explicit expressions for the free energy of antiferromagnetic Heisenberg model with $S = 1$ on a Bravais lattice by means of Popov-Fedotov formalism. Working in a local coordinates and parametrizing a magnetically ordered phase by a magnetic ordering vector $\vec{Q}$, we have derived the general equations of a sublattice magnetization and free energy in unique forms for any Bravais lattice and any magnetically ordered state. For a particular model, one need to find the ordering vectors $\vec{Q}$, the quantities $X(\vec{p}), Y(\vec{p}), W(\vec{p})$ and then one can apply directly the Eqs.(45), (57), (58). It is easy to check that at zero temperature limit it does not matter whether the constraint is treated exactly or average. The magnetization and the free energy are the same, as for the case $S = 1/2$. However, at finite temperature the exact local constraint gives a significant effect. As an example we consider the antiferromagnetic Heisenberg model on D-dimensional hypercubic and triangular lattice with nearest - neighbour bonds.

The Fourier transform of the exchange interactions are given respectively as follows:

$$J_{HC}(\vec{p}) = 2J \sum_{k=1}^{D} \cos \vec{p}_x \tag{73}$$

$$J_{TRL}(\vec{p}) = 2J \left(\cos \vec{p}_x + 2 \cos \frac{\vec{p}_x}{2} \cos \frac{\sqrt{3}}{2}\vec{p}_y\right) \tag{74}$$

Minimizing $J_{HC}(\vec{p})$ and $J_{TRL}(\vec{p})$ with respect to $\vec{p}$, we find that the classical state is described by ordering vectors $Q_{HC} = (\pi, \pi)$ for the D-dimensional hypercubic lattice and $Q_{TRL} = 2\pi\left(\frac{1}{3}, \frac{1}{\sqrt{3}}\right)$ for the triangular lattice. According to the Eqs. (40) and (56), we obtain:

For the hypercubic lattice:

$$\begin{cases} \lambda_{HC} = 2JD \\ X_{HC}(\vec{p}) = -Y_{HC}(\vec{p}) = J_{HC}(\vec{p}) \end{cases} \tag{75}$$

For the triangular lattice:

$$\begin{cases} \lambda_{TRL} = 3J \\ X_{TRL}(\vec{p}) = -\frac{1}{2}Y_{TRL}(\vec{p}) = \frac{3}{2}J(\vec{p}) \end{cases} \tag{76}$$

Subtituting Eqs. (75) and (76) in the Eqs. in Section III, we obtain the magnetization and the free energy for Neel state in the hypercubic lattice and for 120° spin structure in the triangular lattice, taking into account the local constraint condition in one loop approximation. The obtained expressions have a similar structure as for the case $S = 1/2$ [5-7]. Taking limit of zero temperature $T \to 0K$ for the obtained equations, we get the same ground state energy and sublattice magnetization derived in linear spin wave approximation by means of Holstein Primakov representation [7].

The results of this report could be applied for investigating anisotropic Heisenberg models beyond a nearest-neighbour bonds. It would be also interesting to extend the above results for the case of arbitrary quantum spin number S and for a non-Bravais lattice. It is left for a future.

Acknowledgment

This work was financially supported by the National Foundation for Science and Technology Development (NAFOSTED) of Vietnam under Grant No 103.01- 2014.23.

References

[1] Auerbach 1994, *"Interacting electrons and quantum magnetism"*, Springer Verlag
[2] Popov N and Fedotov S A 1988, *Sov. Phys. JETP* **67** 535
[3] Veits O et al 1994, *J. France* **4** 493
[4] Tejima S and Oguchi A 1995, *J. Phys. Soc. Jpn.* **64** 4923
[5] Azakov S, Dilaver M and Oztas 2000, *Int. J. Mod. Phys.* B **14** 13
[6] Dillenschneider R and Richert J 2006, *Eur. Phys. J.* B **49** 187
[7] Pham Thi Thanh Nga and Nguyen Toan Thang 2012, *Comm. in Phys.* **22** 33 (2012),*Comm. in Phys.* **22** 383 (2012) Erratum
[8] Haldane F D 1983, *Phys. Rev. Lett.* **50** 1153
[9] Misguich G and Lhuillier C 2015, in *"Frustrated Spin Systems"*, Ed. by H. T. Diep, World Scientific Singapore, p.299
[10] Miyake S J 1992, *J. Phys. Soc. Jpn.* **91** 938